*"What Do You Care
What Other People Think?"*

'WHAT DO *YOU* CARE WHAT OTHER PEOPLE THINK?'

FURTHER ADVENTURES OF A CURIOUS CHARACTER

RICHARD P. FEYNMAN

as told to Ralph Leighton

HarperCollins*Publishers*

HarperCollins*Publishers*
77–85 Fulham Palace Road,
Hammersmith, London W6 8JB

This paperback edition 1993
3 5 7 9 8 6 4 2

Previously published in paperback by Grafton 1992

First published in Great Britain by
Unwin Hyman Ltd 1989

Copyright © Gwyneth Feynman and Ralph Leighton, 1988

ISBN 0 586 21855 6

Set in Baskerville

Printed in Great Britain by
HarperCollinsManufacturing Glasgow

Contents

Preface

BECAUSE of the appearance of "Surely You're Joking, Mr. Feynman!" a few things need to be explained here.

First, although the central character in this book is the same as before, the "adventures of a curious character" here are different: some are light and some tragic, but most of the time Mr. Feynman is surely *not* joking—although it's often hard to tell.

Second, the stories in this book fit together more loosely than those in "Surely You're Joking . . . ," where they were arranged chronologically to give a semblance of order. (That resulted in some readers getting the mistaken idea that *SYJ* is an autobiography.) My motivation is simple: ever since hearing my first Feynman stories, I have had the powerful desire to share them with others.

Finally, most of these stories were not told at drumming sessions, as before. I will elaborate on this in the brief outline that follows.

Part 1, "A Curious Character," begins by describing the influence of those who most shaped Feynman's personality—his father, Mel, and his first love, Arlene. The first story was adapted from "The Pleasure of Finding Things Out," a BBC program produced by Christopher Sykes. The story of Arlene, from which the title of this book was taken, was painful for Feynman to recount. It was assembled over the past ten years out of pieces from six different stories. When it was finally complete, Feynman was especially fond of this story, and happy to share it with others.

The other Feynman stories in Part 1, although generally lighter in tone, are included here because there won't be a second volume of *SYJ*. Feynman was particularly proud of "It's as Simple as One, Two, Three," which he occasionally thought of writing up as a psychology paper. The letters in the last chapter of Part 1 have been provided courtesy of Gweneth Feynman, Freeman Dyson, and Henry Bethe.

Part 2, "Mr. Feynman Goes to Washington," is, unfortunately, Feynman's last big adventure. The story is particularly long because its content is still timely. (Shorter versions have appeared in *Engineering and Science* and *Physics Today*.) It was not published sooner because Feynman underwent his third and fourth major surgeries—plus radiation, hyperthermia, and other treatments—since serving on the Rogers Commission.

Feynman's decade-long battle against cancer ended on February 15, 1988, two weeks after he taught his last class at Caltech. I decided to include one of his most eloquent and inspirational speeches, "The Value of Science," as an epilogue.

Ralph Leighton
March 1988

"What Do _You_ Care What Other People Think?"

The

Making of a

Scientist

I HAVE a friend who's an artist, and he sometimes takes a view which I don't agree with. He'll hold up a flower and say, "Look how beautiful it is," and I'll agree. But then he'll say, "I, as an artist, can see how beautiful a flower is. But you, as a scientist, take it all apart and it becomes dull." I think he's kind of nutty.

First of all, the beauty that he sees is available to other people—and to me, too, I believe. Although I might not be quite as refined aesthetically as he is, I can appreciate the beauty of a flower. But at the same time, I see much more in the flower than he sees. I can imagine the cells inside, which also have a beauty. There's beauty not just at the dimension of one centimeter; there's also beauty at a smaller dimension.

There are the complicated actions of the cells, and other processes. The fact that the colors in the flower have evolved in order to attract insects to pollinate it is interesting; that means insects can see the colors. That adds a question: does this aesthetic sense we have also exist in lower forms of life? There are all kinds of interesting questions that come from a knowledge of science, which only adds to the excitement and mystery and awe of a flower. It only adds. I don't understand how it subtracts.

I've always been very one-sided about science, and when I was younger

I concentrated almost all my effort on it. In those days I didn't have time, and I didn't have much patience, to learn what's called the humanities. Even though there were humanities courses in the university that you had to take in order to graduate, I tried my best to avoid them. It's only afterwards, when I've gotten older and more relaxed, that I've spread out a little bit. I've learned to draw and I read a little bit, but I'm really still a very one-sided person and I don't know a great deal. I have a limited intelligence and I use it in a particular direction.

Before I was born, my father told my mother, "If it's a boy, he's going to be a scientist."* When I was just a little kid, very small in a highchair, my father brought home a lot of little bathroom tiles—seconds—of different colors. We played with them, my father setting them up vertically on my highchair like dominoes, and I would push one end so they would all go down.

Then after a while, I'd help set them up. Pretty soon, we're setting them up in a more complicated way: two white tiles and a blue tile, two white tiles and a blue tile, and so on. When my mother saw that she said, "Leave the poor child alone. If he wants to put a blue tile, let him put a blue tile."

But my father said, "No, I want to show him what patterns are like and how interesting they are. It's a kind of elementary mathematics." So he started very early to tell me about the world and how interesting it is.

We had the *Encyclopaedia Britannica* at home. When I was a small boy he used to sit me on his lap and read to me from the *Britannica*. We would be reading, say, about dinosaurs. It would be talking about the *Tyrannosaurus rex,* and it would say something like, "This dinosaur is twenty-five feet high and its head is six feet across."

*Richard's younger sister, Joan, has a Ph.D. in physics, in spite of this preconception that only boys are destined to be scientists.

My father would stop reading and say, "Now, let's see what that means. That would mean that if he stood in our front yard, he would be tall enough to put his head through our window up here." (We were on the second floor.) "But his head would be too wide to fit in the window." Everything he read to me he would translate as best he could into some reality.

It was very exciting and very, very interesting to think there were animals of such magnitude—and that they all died out, and that nobody knew why. I wasn't frightened that there would be one coming in my window as a consequence of this. But I learned from my father to translate: everything I read I try to figure out what it really means, what it's really saying.

We used to go to the Catskill Mountains, a place where people from New York City would go in the summer. The fathers would all return to New York to work during the week, and come back only for the weekend. On weekends, my father would take me for walks in the woods and he'd tell me about interesting things that were going on in the woods. When the other mothers saw this, they thought it was wonderful and that the other fathers should take their sons for walks. They tried to work on them but they didn't get anywhere at first. They wanted my father to take all the kids, but he didn't want to because he had a special relationship with me. So it ended up that the other fathers had to take their children for walks the next weekend.

The next Monday, when the fathers were all back at work, we kids were playing in a field. One kid says to me, "See that bird? What kind of bird is that?"

I said, "I haven't the slightest idea what kind of a bird it is."

He says, "It's a brown-throated thrush. Your father doesn't teach you anything!"

But it was the opposite. He had already taught me: "See that bird?" he says. "It's a Spencer's warbler." (I knew he didn't know the real name.) "Well, in Italian, it's a *Chutto Lapittida*. In Portuguese, it's a *Bom da Peida*. In Chinese, it's a *Chung-long-tah*, and in Japanese, it's a *Katano Tekeda*. You can know the name of that bird in all the languages of the world, but when you're finished, you'll know absolutely nothing whatever about the bird. You'll only know about humans in different places, and what they call the bird. So let's look at the bird and see what it's *doing*—that's what counts." (I learned very early the difference between knowing the name of something and knowing something.)

He said, "For example, look: the bird pecks at its feathers all the time. See it walking around, pecking at its feathers?"

"Yeah."

He says, "Why do you think birds peck at their feathers?"

I said, "Well, maybe they mess up their feathers when they fly, so they're pecking them in order to straighten them out."

"All right," he says. "If that were the case, then they would peck a lot just after they've been flying. Then, after they've been on the ground a while, they wouldn't peck so much any more—you know what I mean?"

"Yeah."

He says, "Let's look and see if they peck more just after they land."

It wasn't hard to tell: there was not much difference between the birds that had been walking around a bit and those that had just landed. So I said, "I give up. Why does a bird peck at its feathers?"

"Because there are lice bothering it," he says. "The lice eat flakes of protein that come off its feathers."

He continued, "Each louse has some waxy stuff on its legs, and little mites eat that. The mites don't digest it perfectly, so they emit from their rear ends a sugar-like material, in which bacteria grow."

Finally he says, "So you see, everywhere there's a source of food, there's *some* form of life that finds it."

Now, I knew that it may not have been exactly a louse, that it might not be exactly true that the louse's legs have mites. That story was probably incorrect in *detail*, but what he was telling me was right in *principle*.

Another time, when I was older, he picked a leaf off of a tree. This leaf had a flaw, a thing we never look at much. The leaf was sort of deteriorated; it had a little brown line in the shape of a C, starting somewhere in the middle of the leaf and going out in a curl to the edge.

"Look at this brown line," he says. "It's narrow at the beginning and it's wider as it goes to the edge. What this is, is a fly—a blue fly with yellow eyes and green wings has come and laid an egg on this leaf. Then, when the egg hatches into a maggot (a caterpillar-like thing), it spends its whole life eating this leaf—that's where it gets its food. As it eats along, it leaves behind this brown trail of eaten leaf. As the maggot grows, the trail grows wider until he's grown to full size at the end of the leaf, where he turns into a fly—a blue fly with yellow eyes and green wings—who flies away and lays an egg on another leaf."

Again, I knew that the details weren't precisely correct—it could have even been a beetle—but the idea that he was trying to explain to me was the amusing part of life: the whole thing is just reproduction. No matter how complicated the business is, the main point is to do it again!

Not having experience with many fathers, I didn't realize how remarkable he was. How did he learn the deep principles of science and the love of it, what's behind it, and why it's worth doing? I never really asked him, because I just assumed that those were things that fathers knew.

My father taught me to notice things. One day, I was playing with an "express wagon," a little wagon with a railing around it. It had a ball in it, and when I pulled the wagon, I noticed something about the way the ball moved.

I went to my father and said, "Say, Pop, I noticed something. When I pull the wagon, the ball rolls to the back of the wagon. And when I'm pulling it along and I suddenly stop, the ball rolls to the front of the wagon. Why is that?"

"That, nobody knows," he said. "The general principle is that things which are moving tend to keep on moving, and things which are standing still tend to stand still, unless you push them hard. This tendency is called 'inertia,' but nobody knows why it's true." Now, that's a deep understanding. He didn't just give me the name.

He went on to say, "If you look from the side, you'll see that it's the back of the wagon that you're pulling against the ball, and the ball stands still. As a matter of fact, from the friction it starts to move forward a little bit in relation to the ground. It doesn't move back."

I ran back to the little wagon and set the ball up again and pulled the wagon. Looking sideways, I saw that indeed he was right. Relative to the sidewalk, it moved forward a little bit.

That's the way I was educated by my father, with those kinds of examples and discussions: no pressure—just lovely, interesting discussions. It has motivated me for the rest of my life, and makes me interested in *all* the sciences. (It just happens I do physics better.)

I've been caught, so to speak—like someone who was given something wonderful when he was a child, and he's always looking for it again. I'm always looking, like a child, for the wonders I know I'm going to find—maybe not every time, but every once in a while.

Around that time my cousin, who was three years older, was in high school. He was having considerable difficulty with his algebra, so a tutor would come. I was allowed to sit in a corner while the tutor would try to teach my cousin algebra. I'd hear him talking about x.

I said to my cousin, "What are you trying to do?"

"I'm trying to find out what x is, like in $2x + 7 = 15$."

I say, "You mean 4."

"Yeah, but you did it by arithmetic. You have to do it by algebra."

I learned algebra, fortunately, not by going to school, but by finding my aunt's old schoolbook in the attic, and understanding that the whole idea was to find out what x is—it doesn't make any difference how you do it. For me, there was no such thing as doing it "by arithmetic," or doing it "by algebra." "Doing it by algebra" was a set of rules which, if you followed them blindly, could produce the answer: "subtract 7 from both sides; if you have a multiplier, divide both sides by the multiplier," and so on—a series of steps by which you could get the answer if you didn't understand what you were trying to do. The rules had been invented so that the children who have to study algebra can all pass it. And that's why my cousin was never able to do algebra.

There was a series of math books in our local library which started out with *Arithmetic for the Practical Man.* Then came *Algebra for the Practical Man,* and then *Trigonometry for the Practical Man.* (I learned trigonometry from that, but I soon forgot it again, because I didn't understand it very well.) When I was about thirteen, the library was going to get *Calculus for the Practical Man.* By this time I knew, from reading the encyclopedia, that calculus was an important and interesting subject, and I ought to learn it.

When I finally saw the calculus book at the library, I was very excited. I went to the librarian to check it out, but she looked at me and said, "You're just a child. What are you taking this book out for?"

It was one of the few times in my life I was uncomfortable and I lied. I said it was for my father.

I took the book home and I began to learn calculus from it. I thought it was relatively simple and straightforward. My father started to read it, but he found it confusing and he couldn't understand it. So I tried to explain calculus to him. I didn't know he was so limited, and it bothered me a little bit. It was the first time I realized that I had learned more in some sense than he.

One of the things that my father taught me besides physics—whether it's correct or not—was a disrespect for certain kinds of things. For example, when I was a little boy, and he would sit me on his knee, he'd show me rotogravures in the *New York Times*—that's printed pictures which had just come out in newspapers.

One time we were looking at a picture of the pope and everybody bowing in front of him. My father said, "Now, look at those humans. Here's one human standing here, and all these others are bowing in front of him. Now, what's the difference? This one is the pope"—he hated the pope anyway. He said, "This difference is the hat he's wearing." (If it was a general, it was the epaulets. It was always the costume, the uniform, the position.) "But," he said, "this man has the same problems as everybody else: he eats dinner; he goes to the bathroom. He's a human being." (By the way, my father was in the uniform business, so he knew what the difference is in a man with the uniform off and the uniform on—it was the same man for him.)

He was happy with me, I believe. Once, though, when I came back from MIT (I'd been there a few years), he said to me, "Now that you've become educated about these things, there's one question I've always had that I've never understood very well."

I asked him what it was.

He said, "I understand that when an atom makes a transition from one state to another, it emits a particle of light called a photon."

"That's right," I said.

He says, "Is the photon in the atom ahead of time?"

"No, there's no photon beforehand."

"Well," he says, "where does it come from, then? How does it come out?"

I tried to explain it to him—that photon numbers aren't conserved; they're just created by the motion of the electron—but I couldn't explain it very well. I said, "It's like the sound that I'm making now: it wasn't in me before." (It's not like my little boy, who suddenly announced one day, when he was very young, that he could no longer say a certain word—the word turned out to be "cat"—because his "word bag" had run out of the word. There's no word bag that makes you use up words as they come out; in the same sense, there's no "photon bag" in an atom.)

He was not satisfied with me in that respect. I was never able to explain any of the things that he didn't understand. So he was unsuccessful: he sent me to all these universities in order to find out those things, and he never did find out.

Although my mother didn't know anything about science, she had a great influence on me as well. In particular, she had a wonderful sense of humor, and I learned from her that the highest forms of understanding we can achieve are laughter and human compassion.

"What Do You Care What Other People Think?"

WHEN I was a young fella, about thirteen, I had somehow gotten in with a group of guys who were a little older than I was, and more sophisticated. They knew a lot of different girls, and would go out with them—often to the beach.

One time when we were at the beach, most of the guys had gone out on some jetty with the girls. I was interested in a particular girl a little bit, and sort of thought out loud: "Gee, I think I'd like to take Barbara to the movies . . ."

That's all I had to say, and the guy next to me gets all excited. He runs out onto the rocks and finds her. He pushes her back, all the while saying in a loud voice, "Feynman has something he wants to say to you, Barbara!" It was most embarrassing.

Pretty soon the guys are all standing around me, saying, "Well, *say* it, Feynman!" So I invited her to the movies. It was my first date.

I went home and told my mother about it. She gave me all kinds of advice on how to do this and that. For example, if we take the bus, I'm supposed to get off the bus first, and offer Barbara my hand. Or if we have to walk in the street, I'm supposed to walk on the outside. She even told me what kinds of things to say. She was handing down a cultural tradition to me: the women teach their sons how to treat the next generation of women well.

After dinner, I get all slicked up and go to Barbara's house to call for her. I'm nervous. She isn't ready, of course (it's always like that) so her family has me wait for her in the dining room, where they're eating with friends—a lot of people. They say things like, "Isn't he cute!" and all kinds of other stuff. I didn't feel cute. It was absolutely terrible!

I remember everything about the date. As we walked from her house to the new, little theater in town, we talked about playing the piano. I told her how, when I was younger, they made me learn piano for a while, but after six months I was still playing "Dance of the Daisies," and couldn't stand it any more. You see, I was worried about being a sissy, and to be stuck for weeks playing "Dance of the Daisies" was too much for me, so I quit. I was so sensitive about being a sissy that it even bothered me when my mother sent me to the market to buy some snacks called Peppermint Patties and Toasted Dainties.

We saw the movie, and I walked her back to her home. I complimented her on the nice, pretty gloves she was wearing. Then I said goodnight to her on the doorstep.

Barbara says to me, "Thank you for a very lovely evening."

"You're welcome!" I answered. I felt terrific.

The next time I went out on a date—it was with a different girl—I say goodnight to her, and she says, "Thank you for a very lovely evening."

I didn't feel quite so terrific.

When I said goodnight to the third girl I took out, she's got her mouth open, ready to speak, and I say, "Thank you for a very lovely evening!"

She says, "Thank you—uh—Oh!—Yes—uh, I had a lovely evening, too, thank you!"

One time I was at a party with my beach crowd, and one of the older guys was in the kitchen teaching us how to kiss, using his girlfriend to demonstrate: "You have to have your lips like this, at right angles, so the noses don't collide," and so on. So I go into the living room and find a girl. I'm sitting on the couch with my arm around her, practicing this new art, when suddenly there's all kinds of excitement: "Arlene is coming! Arlene is coming!" I don't know who Arlene is.

Then someone says, "She's here! She's here!"—and everybody stops what they're doing and jumps up to see this queen. Arlene was very pretty, and I could see why she had all this admiration—it was well deserved—but I didn't believe in this undemocratic business of changing what you're doing just because the queen is coming in.

So, while everybody's going over to see Arlene, I'm still sitting there on the couch with my girl.

(Arlene told me later, after I had gotten to know her, that she remembered that party with all the nice people—except for one guy who was over in the corner on the couch smooching with a girl. What she didn't know was that two minutes before, all the others were doin' it too!)

The first time I ever said anything to Arlene was at a dance. She was very popular, and everybody was cutting in and dancing with her. I remember thinking I'd like to dance with her, too, and trying to decide when to cut in. I always had trouble with that problem: first of all, when she's over on the other side of the dance floor dancing with some guy, it's too complicated—so you wait until they come closer. Then when she's near you, you think, "Well, no, this isn't the kind of music I'm good at dancing to." So you wait for another type of music. When the music changes to something you like, you sort of step forward—at least you *think* you step forward to cut in—when some other guy cuts in just in front of you. So now you have to wait a few minutes because it's impolite to cut in too soon after someone else has. And by the time a few minutes have passed, they're over at the other side of the dance floor again, or the music has changed again, or whatever!

After a certain amount of this stalling and fooling around, I finally mutter something about wanting to dance with Arlene. One of the guys I was hanging around with overhears me and makes a big announcement to the other guys: "Hey, listen to this, guys; Feynman wants to dance with Arlene!" Soon one of them is dancing with Arlene and they dance over towards the rest of us. The others push me out onto the dance floor and I finally "cut in." You can see the condition I was in by my first words to her, which were an honest question: "How does it feel to be so popular?" We only danced a few minutes before somebody else cut in.

My friends and I had taken dancing lessons, although none of us would ever admit it. In those depression days, a friend of my mother was trying to make a living by teaching dancing in the evening, in an upstairs dance studio. There was a back door to the place, and she arranged it so the young men could come up through the back way without being seen.

Every once in a while there would be a social dance at this lady's studio. I didn't have the nerve to test this analysis, but it seemed to me that the girls had a much harder time than the boys did. In those days, girls couldn't ask to cut in and dance with boys; it wasn't "proper." So the girls who weren't very pretty would sit for hours at the side, just sad as hell.

I thought, "The guys have it easy: they're free to cut in whenever they want." But it wasn't easy. You're "free," but you haven't got the guts, or the sense, or whatever it takes to relax and enjoy dancing. Instead, you tie yourself in knots worrying about cutting in or inviting a girl to dance with you.

For example, if you saw a girl who was not dancing, who you thought you'd like to dance with, you might think, "Good! Now at least I've got a chance!" But it was usually very difficult: often the girl would say, "No, thank you, I'm tired. I think I'll sit this one out." So you go away somewhat defeated—but not completely, because maybe she really *is* tired—when you turn around and some other guy comes up to her, and there she is, dancing with him! Maybe this guy is her boyfriend and she knew he was coming over, or maybe she didn't like the way you look, or maybe something else. It was always so complicated for such a simple matter.

One time I decided to invite Arlene to one of these dances. It was the first time I took her out. My best friends were also at the dance; my mother had invited them, to get more customers for her friend's dance studio. These guys were contemporaries of mine, guys my own age from school. Harold Gast and David Leff were literary types, while Robert Stapler was a scientific type. We would spend a lot of time together after school, going on walks and discussing this and that.

Anyway, my best friends were at the dance, and as soon as they saw me with Arlene, they called me into the cloakroom and said, "Now listen, Feynman, we want you to understand that *we* understand that Arlene is *your* girl tonight, and we're not gonna bother you with her. She's out of bounds for us," and so on. But before long, there was cutting in and competition coming from precisely these guys! I learned the meaning of Shakespeare's phrase "Methinks thou dost protest too much."

You must appreciate what I was like then. I was a very shy character, always feeling uncomfortable because everybody was stronger than I, and always afraid I would look like a sissy. Everybody else played baseball; everybody else did all kinds of athletic things. If there was a game somewhere, and a ball would come rolling across the road, I would be petrified that maybe I'd have to pick it up and throw it back—because if I threw it, it would be about a radian off the correct direction, and not anywhere near the distance! And then everybody would laugh. It was terrible, and I was very unhappy about it.

One time I was invited to a party at Arlene's house. Everybody was there because Arlene was the most popular girl around: she was number one, the nicest girl, and everybody liked her. Well, I'm sitting in a big armchair with nothing to do, when Arlene comes over and sits on the arm of the chair to talk to me. That was the beginning of the feeling, "Oh, boy! The world is just wonderful now! Somebody I like has paid attention to me!"

In those days, in Far Rockaway, there was a youth center for Jewish kids at the temple. It was a big club that had many activities. There was a writers group that wrote stories and would read them to each other; there was a drama group that put on plays; there was a science group, and an art group. I had no interest in any subject except science, but Arlene was in the art group, so I joined it too. I struggled with the art business—learning how to make plaster casts of the face and so on (which I used much later in life, it turned out)—just so I could be in the same group with Arlene.

But Arlene had a boyfriend named Jerome in the group, so there was no chance for me. I just hovered around in the background.

One time, when I wasn't there, somebody nominated me for president of the youth center. The elders began getting nervous, because I was an avowed atheist by that time.

I had been brought up in the Jewish religion—my family went to the temple every Friday, I was sent to what we called "Sunday school," and I even studied Hebrew for a while—but at the same time, my father was telling me about the world. When I would hear the rabbi tell about some miracle such as a bush whose leaves were shaking but there wasn't any wind, I would try to fit the miracle into the real world and explain it in terms of natural phenomena.

Some miracles were harder than others to understand.

The one about the leaves was easy. When I was walking to school, I heard a little noise: although the wind was hardly noticeable, the leaves of a bush were wiggling a little bit because they were in just the right position to make a kind of resonance. And I thought, "Aha! This is a good explanation for Elijah's vision of the quaking bush!"

But there were some miracles I never did figure out. For instance, there was a story in which Moses throws down his staff and it turns into a snake. I couldn't figure out what the witnesses saw that made them think his staff was a snake.

If I had thought back to when I was much younger, the Santa Claus story could have provided a clue for me. But it didn't hit me hard enough at the time to produce the possibility that I should doubt the truth of stories that don't fit with nature. When I found out that Santa Claus wasn't real, I wasn't upset; rather, I was relieved that there was a much simpler phenomenon to explain how so many children all over the world got presents on the same night! The story had been getting pretty complicated—it was getting out of hand.

Santa Claus was a particular custom we celebrated in our family, and it wasn't very serious. But the miracles I was hearing about were connected with real things: there was the temple, where people would go every week; there was the Sunday school, where rabbis taught children about miracles; it was much more of a dramatic thing. Santa Claus didn't involve big institutions like the temple, which I knew were real.

So all the time I was going to the Sunday school, I was believing everything and having trouble putting it together. But of course, ultimately, it had to come to a crisis, sooner or later.

The actual crisis came when I was eleven or twelve. The rabbi was telling us a story about the Spanish Inquisition, in which the Jews suffered terrible tortures. He told us about a particular individual whose name was Ruth, exactly what she was supposed to have done, what the arguments were in her favor and against her—the whole thing, as if it had all been documented by a court reporter. And I was just an innocent kid, listening to all this stuff and believing it was a true commentary, because the rabbi had never indicated otherwise.

At the end, the rabbi described how Ruth was dying in prison: "And she thought, while she was dying"—blah, blah.

That was a shock to me. After the lesson was over, I went up to him and said, "How did they know what she thought when she was dying?"

He says, "Well, of course, in order to explain more vividly how the Jews suffered, we made up the story of Ruth. It wasn't a real individual."

That was too much for me. I felt terribly deceived: I wanted the straight story—not fixed up by somebody else—so I could decide for myself what it meant. But it was difficult for me to argue with adults. All I could do was get tears in my eyes. I started to cry, I was so upset.

He said, "What's the matter?"

I tried to explain. "I've been listening to all these stories, and now I don't know, of all the things you told me, which were true, and which were not true! I don't know what to do with everything that I've learned!" I was trying to explain that I was losing everything at the moment, because I was no longer sure of the data, so to speak. Here I had been struggling to understand all these miracles, and now—well, it solved a lot of miracles, all right! But I was unhappy.

The rabbi said, "If it is so traumatic for you, why do you come to Sunday school?"

"Because my parents make me."

I never talked to my parents about it, and I never found out whether the rabbi communicated with them or not, but my parents never made me go again. And it was just before I was supposed to get confirmed as a believer.

Anyway, that crisis resolved my difficulty rather rapidly, in favor of the theory that all the miracles were stories made up to help people understand things "more vividly," even if they conflicted with natural phenomena. But I thought nature itself was so interesting that I didn't want it distorted like that. And so I gradually came to disbelieve the whole religion.

Anyway, the Jewish elders had organized this club with all its activities not just to get us kids off the street, but to get us interested in the Jewish way of life. So to have someone like me elected as president would have made them very embarrassed. To our mutual relief I wasn't elected, but the center eventually failed anyway—it was on its way out when I was nominated, and had I been elected, I surely would have been blamed for its demise.

One day Arlene told me Jerome isn't her boyfriend anymore. She's not tied up with him. That was a big excitement for me, the beginning of *hope!* She invited me over to her house, at 154 Westminster Avenue in nearby Cedarhurst.

When I went to her house that time, it was dark and the porch wasn't lit. I couldn't see the numbers. Not wanting to disturb anyone by asking if it was the right house, I crawled up, quietly, and felt the numbers on the door: 154.

Arlene was having trouble with her homework in philosophy class. "We're studying Descartes," she said. "He starts out with 'Cogito, ergo sum'—'I think, therefore I am'—and ends up proving the existence of God."

"Impossible!" I said, without stopping to think that I was doubting the great Descartes. (It was a reaction I learned from my father: have no respect *whatsoever* for authority; forget who said it and instead look at what he starts with, where he ends up, and ask yourself, "Is it reasonable?") I said, "How can you deduce one from the other?"

"I don't know," she said.

"Well, let's look it over," I said. "What's the argument?"

So we look it over, and we see that Descartes' statement "Cogito, ergo sum" is supposed to mean that there is one thing that cannot be doubted—doubt itself. "Why doesn't he just say it straight?" I complained. "He just means somehow or other that he has one fact that he knows."

Then it goes on and says things like, "I can only imagine imperfect thoughts, but imperfect can only be understood as referent to the perfect. Hence the perfect must exist somewhere." (He's workin' his way towards God now.)

"Not at all!" I say. "In science you can talk about relative degrees of approximation without having a perfect theory. I don't know what this is all about. I think it's a bunch of baloney."

Arlene understood me. She understood, when she looked at it, that no matter how impressive and important this philosophy stuff was supposed to be, it could be taken lightly—you could just think about the words, instead of worrying about the fact that Descartes said it. "Well, I guess it's okay to take the other side," she said. "My teacher keeps telling us, 'There are two sides to every question, just like there are two sides to every piece of paper.'"

"There's two sides to that, too," I said.

"What do you mean?"

I had read about the Möbius strip in the *Britannica*, my wonderful *Britannica!* In those days, things like the Möbius strip weren't so well known to everybody, but they were just as understandable as they are to kids today. The existence of such a surface was so real: it wasn't a wishy-washy political question, or anything that you needed history to understand. Reading about those things was like being way off in a wonderful world that nobody knows about, and you're getting a kick not only from the delight of learning the stuff itself, but also from making yourself unique.

I got a strip of paper, put a half twist in it, and made it into a loop. Arlene was delighted.

The next day, in class, she lay in wait for her teacher. Sure enough, he holds up a piece of paper and says, "There are two sides to every question, just like there are two sides to every piece of paper." Arlene holds up her own strip of paper—with a half twist in it—and says, "Sir, there are even two sides to *that* question: there's paper with only one side!" The teacher and the class got all excited, and Arlene got such a kick out of showing them the Möbius strip that I think she paid more attention to me after that on account of it.

But after Jerome, I had a new competitor—my "good friend" Harold Gast. Arlene was always making up her mind one way or the other. When it came time for graduation, she went with Harold to the senior prom, but sat with my parents for the graduation ceremony.

I was the best in science, the best in mathematics, the best in physics, and the best in chemistry, so I was going up to the stage and receiving honors many times at the ceremony. Harold was the best in English and the best in history, and had written the school play, so that was very impressive.

I was terrible in English. I couldn't stand the subject. It seemed to me ridiculous to worry about whether you spelled something wrong or not, because English spelling is just a human convention—it has nothing to do with anything *real*, anything from nature. Any word can be spelled just as well a different way. I was impatient with all this English stuff.

There was a series of exams called the Regents, which the state of New York gave to every high school student. A few months before, when we all were taking the Regents examination in English, Harold and the other literary friend of mine, David Leff—the editor of the school newspaper—asked me which books I had chosen to write about. David had chosen something with profound social implications by Sinclair Lewis, and Harold had picked some playwright. I said I chose *Treasure Island* because we had that book in first-year English, and told them what I wrote.

They laughed. "Boy, are *you* gonna flunk, saying such simple stuff about such a simple book!"

There was also a list of questions for an essay. The one I chose was "The Importance of Science in Aviation." I thought, "What a dumb question! The importance of science in aviation is obvious!"

I was about to write a simple theme about this dumb question when I remembered that my literary friends were always "throwing the bull"—building up their sentences to sound complex and sophisticated. I decided to try it, just for the hell of it. I thought, "If the Regents are so silly as to have a subject like the importance of science in aviation, I'm gonna *do* that."

So I wrote stuff like, "Aeronautical science is important in the analysis of the eddies, vortices, and whirlpools formed in the atmosphere behind the aircraft . . ."—I knew that eddies, vortices, and whirlpools are the same thing, but mentioning them three different ways *sounds* better! That was the only thing I would not have ordinarily done on the test.

The teacher who corrected my examination must have been impressed by eddies, vortices, and whirlpools, because I got a 91 on the exam—while my literary friends, who chose topics the English teachers could more easily take issue with, both got 88.

That year a new rule came out: if you got 90 or better on a Regents examination, you automatically got honors in that subject at graduation! So while the playwright and the editor of the school newspaper had to stay in their seats, this illiterate fool physics student was called to go up to the stage once again and receive honors in English!

After the graduation ceremony, Arlene was in the hall with my parents and Harold's parents when the head of the math department came over. He was a very strong man—he was also the school disciplinarian—a tall, dominating fellow. Mrs. Gast says to him, "Hello, Dr. Augsberry. I'm Harold Gast's mother. And this is Mrs. Feynman . . ."

He completely ignores Mrs. Gast and immediately turns to my mother. "Mrs. Feynman, I want to impress upon you that a young man like your son comes along only very rarely. The state should support a man of such talent. You must be *sure* that he goes to college, the best college you can afford!" He was concerned that my parents might not be planning to send me to college, for in those days lots of kids had to get a job immediately after graduation to help support the family.

That in fact happened to my friend Robert. He had a lab, too, and taught me all about lenses and optics. (One day he had an accident in his lab. He was opening carbolic acid and the bottle jerked, spilling some acid on his face. He went to the doctor and had bandages put on for a few weeks. The funny thing was, when they took the bandages off his skin was smooth underneath, nicer than it had been before—there were many fewer blemishes. I've since found out that there was, for a while, some kind of a beauty treatment using carbolic acid in a more dilute form.) Robert's mother was poor, and he had to go to work right away to support her, so he couldn't continue his interest in the sciences.

Anyway, my mother reassured Dr. Augsberry: "We're saving money as best we can, and we're trying to send him to Columbia or MIT." And Arlene was listening to all this, so after that I was a little bit ahead.

Arlene was a wonderful girl. She was the editor of the newspaper at Nassau County Lawrence High School; she played the piano beautifully, and was very artistic. She made some decorations for our house, like the parrot on the inside of our closet. As time went on, and our family got to know her better, she would go to the woods to paint with my father, who had taken up painting in later life, as many people do.

Arlene and I began to mold each other's personality. She lived in a family that was very polite, and was very sensitive to other people's feelings. She taught me to be more sensitive to those kinds of things, too. On the other hand, her family felt that "white lies" were okay.

I thought one should have the attitude of "What do *you* care what other people think!" I said, "We should listen to other people's opinions and take them into account. Then, if they don't make sense and we think they're wrong, then that's that!"

Arlene caught on to the idea right away. It was easy to talk her into thinking that in our relationship, we must be very honest with each other and say everything straight, with absolute frankness. It worked very well, and we became very much in love—a love like no other love that I know of.

After that summer I went away to college at MIT. (I couldn't go to Columbia because of the Jewish quota.*) I began getting letters from my friends that said things like, "You should see how Arlene is going out with Harold," or "She's doing this and she's doing that, while you're all alone up there in Boston." Well, I was taking out girls in Boston, but they didn't mean a thing to me, and I knew the same was true with Arlene.

When summer came, I stayed in Boston for a summer job, and worked on measuring friction. The Chrysler Company had developed a new method of polishing to get a

*Note for foreign readers: the quota system was a discriminatory practice of limiting the number of places in a university available to students of Jewish background.

super finish, and we were supposed to measure how much better it was. (It turned out that the "super finish" was not significantly better.)

Anyway, Arlene found a way to be near me. She found a summer job in Scituate, about twenty miles away, taking care of children. But my father was concerned that I would become too involved with Arlene and get off the track of my studies, so he talked her out of it—or talked me out of it (I can't remember). Those days were very, very different from now. In those days, you had to go all the way up in your career before marrying.

I was able to see Arlene only a few times that summer, but we promised each other we would marry after I finished school. I had known her for six years by that time. I'm a little tongue-tied trying to describe to you how much our love for each other developed, but we were sure we were right for each other.

After I graduated from MIT I went to Princeton, and I would go home on vacations to see Arlene. One time when I went to see her, Arlene had developed a bump on one side of her neck. She was a very beautiful girl, so it worried her a little bit, but it didn't hurt, so she figured it wasn't too serious. She went to her uncle, who was a doctor. He told her to rub it with omega oil.

Then, sometime later, the bump began to change. It got bigger—or maybe it was smaller—and she got a fever. The fever got worse, so the family doctor decided Arlene should go to the hospital. She was told she had typhoid fever. Right away, as I still do today, I looked up the disease in medical books and read all about it.

When I went to see Arlene in the hospital, she was in quarantine—we had to put on special gowns when we entered her room, and so on. The doctor was there, so I asked him how the Wydell test came out—it was an absolute test for typhoid fever that involved checking for bacteria in the feces. He said, "It was negative."

"What? How can that *be!*" I said. "Why all these gowns, when you can't even find the bacteria in an experiment? Maybe she doesn't have typhoid fever!"

The result of that was that the doctor talked to Arlene's parents, who told me not to interfere. "After all, he's the doctor. You're only her fiancé."

I've found out since that such people don't know what they're doing, and get insulted when you make some suggestion or criticism. I realize that *now,* but I wish I had been much stronger then and told her parents that the doctor was an idiot—which he was—and didn't know what he was doing. But as it was, her parents were in charge of it.

Anyway, after a little while, Arlene got better, apparently: the swelling went down and the fever went away. But after some weeks the swelling started again, and this time she went to another doctor. This guy feels under her armpits and in her groin, and so on, and notices there's swelling in those places, too. He says the problem is in her lymphatic glands, but he doesn't yet know what the specific disease is. He will consult with other doctors.

As soon as I hear about it I go down to the library at Princeton and look up lymphatic diseases, and find "Swelling of the Lymphatic Glands. (1) Tuberculosis of the lymphatic glands. This is very easy to diagnose . . ."—so I figure this isn't what Arlene has, because the doctors are having trouble trying to figure it out.

I start reading about some other diseases: lymphodenema, lymphodenoma, Hodgkin's disease, all kinds of other things; they're all cancers of one crazy form or another. The only difference between lymphodenema and lymphodenoma was, as far as I could make out by reading it very carefully, that if the patient dies, it's lymphodenoma; if the patient survives—at least for a while—then it's lymphodenema.

At any rate, I read through all the lymphatic diseases,

and decided that the most likely possibility was that Arlene had an incurable disease. Then I half smiled to myself, thinking, "I bet everybody who reads through a medical book thinks they have a fatal disease." And yet, after reading everything very carefully, I couldn't find any other possibility. It was serious.

Then I went to the weekly tea at Palmer Hall, and found myself talking to the mathematicians just as I always did, even though I had just found out that Arlene probably had a fatal disease. It was very strange—like having two minds.

When I went to visit her, I told Arlene the joke about the people who don't know any medicine reading the medical book and always assuming they have a fatal disease. But I also told her I thought we were in great difficulty, and that the best I could figure out was that she had an incurable disease. We discussed the various diseases, and I told her what each one was like.

One of the diseases I told Arlene about was Hodgkin's disease. When she next saw her doctor, she asked him about it: "Could it be Hodgkin's disease?"

He said, "Well, yes, that's a possibility."

When she went to the county hospital, the doctor wrote the following diagnosis: "Hodgkin's disease—?" So I realized that the doctor didn't know any more than I did about this problem.

The county hospital gave Arlene all sorts of tests and X-ray treatments for this "Hodgkin's disease—?" and there were special meetings to discuss this peculiar case. I remember waiting for her outside, in the hall. When the meeting was over, the nurse wheeled her out in a wheelchair. All of a sudden a little guy comes running out of the meeting room and catches up with us. "Tell me," he says, out of breath, "do you spit up blood? Have you ever coughed up blood?"

The nurse says, "Go away! Go away! What kind of

thing is that to ask of a patient!"—and brushes him away.
Then she turned to us and said, "That man is a doctor from
the neighborhood who comes to the meetings and is always
making trouble. That's not the kind of thing to ask of a
patient!"

I didn't catch on. The doctor was checking a certain
possibility, and if I had been smart, I would have asked him
what it was.

Finally, after a lot of discussion, a doctor at the hospital
tells me they figure the most likely possibility is Hodgkin's
disease. He says, "There will be some periods of improve-
ment, and some periods in the hospital. It will be on and
off, getting gradually worse. There's no way to reverse it
entirely. It's fatal after a few years."

"I'm sorry to hear that," I say. "I'll tell her what you
said."

"No, no!" says the doctor. "We don't want to upset the
patient. "We're going to tell her it's glandular fever."

"No, no!" I reply. "We've already discussed the possibility of Hodgkin's disease. I know she can adjust to it."

"Her parents don't want her to know. You had better
talk to them first."

At home, everybody worked on me: my parents, my
two aunts, our family doctor; they were all on me, saying
I'm a very foolish young man who doesn't realize what pain
he's going to bring to this wonderful girl by telling her she
has a fatal disease. "How can you do such a terrible thing?"
they asked, in horror.

"Because we have made a pact that we must speak
honestly with each other and look at everything directly.
There's no use fooling around. She's gonna ask me what
she's got, and I cannot lie to her!"

"Oh, that's childish!" they said—blah, blah, blah. Everybody kept working on me, and said I was wrong. I
thought I was definitely right, because I had already talked
to Arlene about the disease and knew she could face it—

that telling her the truth was the right way to handle it.

But finally, my little sister comes up to me—she was eleven or twelve then—with tears running down her face. She beats me on the chest, telling me that Arlene is such a wonderful girl, and that I'm such a foolish, stubborn brother. I couldn't take it any more. That broke me down.

So I wrote Arlene a goodbye love letter, figuring that if she ever found out the truth after I had told her it was glandular fever, we would be through. I carried the letter with me all the time.

The gods never make it easy; they always make it harder. I go to the hospital to see Arlene—having made this decision—and there she is, sitting up in bed, surrounded by her parents, somewhat distraught. When she sees me, her face lights up and she says, "Now I know how valuable it is that we tell each other the truth!" Nodding at her parents, she continues, "They're telling me I have glandular fever, and I'm not sure whether I believe them or not. Tell me, Richard, do I have Hodgkin's disease or glandular fever?"

"You have glandular fever," I said, and I died inside. It was terrible—just terrible!

Her reaction was completely simple: "Oh! Fine! Then I believe them." Because we had built up so much trust in each other, she was completely relieved. Everything was solved, and all was very nice.

She got a little bit better, and went home for a while. About a week later, I get a telephone call. "Richard," she says, "I want to talk to you. Come on over."

"Okay." I made sure I still had the letter with me. I could tell something was the matter.

I go upstairs to her room, and she says, "Sit down." I sit down on the end of her bed. "All right, now tell me," she says, "do I have glandular fever or Hodgkin's disease?"

"You have Hodgkin's disease." And I reached for the letter.

"God!" she says. "They must have put you through hell!"

I had just told her she has a fatal disease, and was admitting that I had lied to her as well, and what does she think of? She's worried about *me!* I was terribly ashamed of myself. I gave Arlene the letter.

"You should have stuck by it. We know what we're doing; we are right!"

"I'm sorry. I feel awful."

"I understand, Richard. Just don't do it again."

You see, she was in bed upstairs, and did something she used to do when she was little: she tiptoed out of bed and crawled down the stairs a little bit to listen to what people were doing downstairs. She heard her mother crying a lot, and went back to bed thinking, "If I have glandular fever, why is Mother crying so much? But Richard said I had glandular fever, so it must be right!"

Later she thought, "Could *Richard* have lied to me?" and began to wonder how that might be possible. She concluded that, incredible as it sounded, somebody might have put me through a wringer of some sort.

She was so good at facing difficult situations that she went on to the next problem. "Okay," she says, "I have Hodgkin's disease. What are we going to do now?"

I had a scholarship at Princeton, and they wouldn't let me keep it if I got married. We knew what the disease was like: sometimes it would get better for some months, and Arlene could be at home, and then she would have to be in the hospital for some months—back and forth for two years, perhaps.

So I figure, although I'm in the middle of trying to get my Ph.D., I could get a job at the Bell Telephone Laboratories doing research—it was a very good place to work—and we could get a little apartment in Queens that wasn't too far from the hospital or Bell Labs. We could get married in a few months, in New York. We worked everything out that afternoon.

For some months now Arlene's doctors had wanted to take a biopsy of the swelling on her neck, but her parents didn't want it done—they didn't want to "bother the poor sick girl." But with new resolve, I kept working on them, explaining that it's important to get as much information as possible. With Arlene's help, I finally convinced her parents.

A few days later, Arlene telephones me and says, "They got a report from the biopsy."

"Yeah? Is it good or bad?"

"I don't know. Come over and let's talk about it."

When I got to her house, she showed me the report. It said, "Biopsy shows tuberculosis of the lymphatic gland."

That really got me. I mean, that was the first goddamn thing on the list! I passed it by, because the book said it was easy to diagnose, and because the doctors were having so much trouble trying to figure out what it was. I assumed they had checked the obvious case. And it *was* the obvious case: the man who had come running out of the meeting room asking "Do you spit up blood?" had the right idea. He knew what it probably was!

I felt like a jerk, because I had passed over the obvious possibility by using circumstantial evidence—which isn't any good—and by assuming the doctors were more intelligent than they were. Otherwise, I would have suggested it right off, and perhaps the doctor would have diagnosed Arlene's disease way back then as "tuberculosis of the lymphatic gland—?" I was a dope. I've learned, since then.

Anyway, Arlene says, "So I might live as long as seven years. I may even get better."

"So what do you mean, you don't know if it's good or bad?"

"Well, now we won't be able to get married until later."

Knowing that she only had two more years to live, we had solved things so perfectly, from her point of view, that she was disturbed to discover she'd live longer! But it didn't take me long to convince her it was a better circumstance.

So we knew we could face things together, from then on. After going through that, we had no difficulty facing any other problem.

When the war came, I was recruited to work on the Manhattan Project at Princeton, where I was finishing up my degree. A few months later, as soon as I got my degree, I announced to my family that I wanted to get married.

My father was horrified, because from the earliest times, as he saw me develop, he thought I would be happy as a scientist. He thought it was still too early to marry—it would interfere with my career. He also had this crazy idea: if a guy was in some difficulty, he used to always say, "Cherchez la femme"—look for the woman behind it. He felt that women were the great danger to a man, that a man always has to watch out and be tough about women. And when he sees me marrying a girl with tuberculosis, he thinks of the possibility that I'm going to get sick, too.

My whole family was worried about that—aunts, uncles, everyone. They brought the family doctor over to our house. He tried to explain to me that tuberculosis is a dangerous disease, and that I'm bound to get it.

I said, "Just tell me how it's transmitted, and we'll figure it out." We were already very, very careful: we knew we must not kiss, because there's a lot of bacteria in the mouth.

Then they very carefully explained to me that when I had promised to marry Arlene, I didn't know the situation. Everybody would understand that I didn't know the situation then, and that it didn't represent a real promise.

I never had that feeling, that crazy idea that they had, that I was getting married because I had promised it. I hadn't even *thought* of that. It wasn't a question of having promised anything; we had stalled around, not getting a piece of paper and not being formally married, but we were in love, and were already married, emotionally.

I said, "Would it be sensible for a husband who learns that his wife has tuberculosis to leave her?"

Only my aunt who ran the hotel thought maybe it would be all right for us to get married. Everybody else was still against it. But this time, since my family had given me this kind of advice before and it had been so wrong, I was in a much stronger position. It was very easy to resist and to just proceed. So there was no problem, really. Although it was a similar circumstance, they weren't going to convince me of anything any more. Arlene and I knew we were right in what we were doing.

Arlene and I worked everything out. There was a hospital in New Jersey just south of Fort Dix where she could stay while I was at Princeton. It was a charity hospital— Deborah was the name of it—supported by the Women's Garment Workers Union of New York. Arlene wasn't a garment worker, but it didn't make any difference. And I was just a young fella working on this project for the government, and the pay was very low. But this way I could take care of her, at last.

We decided to get married on the way to Deborah Hospital. I went to Princeton to pick up a car—Bill Woodward, one of the graduate students there, lent me his station wagon. I fixed it up like a little ambulance, with a mattress and sheets in the back, so Arlene could lie down in case she got tired. Although this was one of the periods when the disease was apparently not so bad and she was at home, Arlene had been in the county hospital a lot, and she was a little weak.

I drove up to Cedarhurst and picked up my bride. Arlene's family waved goodbye, and off we went. We crossed Queens and Brooklyn, then went to Staten Island on the ferry—that was our romantic boat ride—and drove to the city hall for the borough of Richmond to get married.

We went up the stairs, slowly, into the office. The guy there was very nice. He did everything right away. He said, "You don't have any witnesses," so he called the bookkeeper and an accountant from another room, and we were married according to the laws of the state of New York. Then we were very happy, and we smiled at each other, holding hands.

The bookkeeper says to me, "You're married now. You should kiss the bride!"

So the bashful character kissed his bride lightly on the cheek.

I gave everyone a tip and we thanked them very much. We got back in the car, and drove to Deborah Hospital.

Every weekend I'd go down from Princeton to visit Arlene. One time the bus was late, and I couldn't get into the hospital. There weren't any hotels nearby, but I had my old sheepskin coat on (so I was warm enough), and I looked for an empty lot to sleep in. I was a little worried what it might look like in the morning when people looked out of their windows, so I found a place that was far enough away from houses.

The next morning I woke up and discovered I'd been sleeping in a garbage dump—a landfill! I felt foolish, and laughed.

Arlene's doctor was very nice, but he would get upset when I brought in a war bond for $18 every month. He could see we didn't have much money, and kept insisting we shouldn't contribute to the hospital, but I did it anyway.

One time, at Princeton, I received a box of pencils in the mail. They were dark green, and in gold letters were the words "RICHARD DARLING, I LOVE YOU! PUTSY." It was Arlene (I called her Putsy).

Well, that was nice, and I love her, too, but—you know

how you absentmindedly drop pencils around: you're showing Professor Wigner a formula, or something, and leave the pencil on his desk.

In those days we didn't have extra stuff, so I didn't want to waste the pencils. I got a razor blade from the bathroom and cut off the writing on one of them to see if I could use them.

The next morning, I get a letter in the mail. It starts out, "WHAT'S THE IDEA OF TRYING TO CUT THE NAME OFF THE PENCILS?"

It continues: "Aren't you proud of the fact that I love you?" Then: "WHAT DO YOU CARE WHAT OTHER PEOPLE THINK?"

Then came poetry: "If you're ashamed of me, dah dah, then Pecans to you! Pecans to you!" The next verse was the same kind of stuff, with the last line, "Almonds to you! Almonds to you!" Each one was "Nuts to you!" in a different form.

So I had to use the pencils with the names on them. What else could I do?

It wasn't long before I had to go to Los Alamos. Robert Oppenheimer, who was in charge of the project, arranged for Arlene to stay in the nearest hospital, in Albuquerque, about a hundred miles away. I had time off every weekend to see her, so I would hitchhike down on a Saturday, see Arlene in the afternoon, and stay overnight in a hotel there in Albuquerque. Then on Sunday morning I would see Arlene again, and hitchhike back to Los Alamos in the afternoon.

During the week I would often get letters from her. Some of them, like the one written on a jigsaw-puzzle blank and then taken apart and sent in a sack, resulted in little notes from the army censor, such as "Please tell your wife we don't have time to play games around here." I didn't tell her anything. I *liked* her to play games—even though she often put me in various uncomfortable but amusing conditions from which I could not escape.

One time, near the beginning of May, newspapers mysteriously appeared in almost everybody's mailbox at Los Alamos. The whole damn place was full of them— hundreds of newspapers. You know the kind—you open it up and there's this headline screaming in thick letters across the front page: ENTIRE NATION CELEBRATES BIRTHDAY OF R.P. FEYNMAN!

Arlene was playing her game with the world. She had a lot of time to think. She would read magazines, and send away for this and that. She was always cooking up something. (She must have got help with the names from Nick Metropolis or one of the other guys at Los Alamos who would often visit her.) Arlene was in her room, but she was in the world, writing me crazy letters and sending away for all kinds of stuff.

One time she sent me a big catalog of kitchen equipment—the kind you need for enormous institutions like prisons, which have a lot of people in them. It showed everything from blowers and hoods for stoves to huge pots and pans. So I'm thinking, "What the hell is this?"

It reminded me of the time I was up at MIT and Arlene sent me a catalog describing huge boats, from warships to ocean liners—great big boats. I wrote to her: "What's the idea?"

She writes back: "I just thought that maybe, when we get married, we could buy a boat."

I write, "Are you crazy? It's all out of proportion!"

Then another catalog comes: it's for big yachts—forty-foot schooners and stuff like that—for very rich people. She writes, "Since you said no to the other boats, maybe we could get one of these."

I write, "Look: you're way out of scale!"

Soon another catalog comes: it's for various kinds of motor boats—Chriscraft this and that.

I write, "Too expensive!"

Finally, I get a note: "This is your last chance, Richard. You're always saying no." It turns out a friend of hers has a rowboat she wants to sell for $15—a used rowboat—and maybe we could buy it so we could row around in the water next summer.

So, yes. I mean, how can you say no after all that?

Well, I'm still trying to figure out what this big catalog for institutional kitchen equipment is leading to, when another catalog comes: it's for hotels and restaurants—supplies for small and medium-sized hotels and restaurants. Then a few days later, a catalog for the kitchen in your new home comes.

When I go down to Albuquerque the next Saturday, I find out what it's all about. There's a little charcoal broiler in her room—she's bought it through the mail from Sears. It's about eighteen inches across, with little legs.

"I thought we could have steaks," Arlene says.

"How the hell can we use it in the room, here, with all the smoke and everything?"

"Oh, no," she says. "All you have to do is take it out on the lawn. Then you can cook us steaks every Sunday."

The hospital was right on Route 66, the main road across the United States! "I can't do that," I said. "I mean, with all the cars and trucks going by, all the people on the sidewalk walking back and forth, I can't just go out there and start cookin' steaks on the lawn!"

"What do *you* care what other people think?" (Arlene *tortured* me with that!) "Okay," she says, opening a drawer, "we'll compromise: you don't have to wear the chef's hat and the gloves."

She holds up a hat—it's a real chef's hat—and gloves. Then she says, "Try on the apron," as she unfolds it. It has something silly written across it, like "BAR-B-Q KING," or something.

"Okay, okay!" I say, horrified. "I'll cook the steaks on

the lawn!" So every Saturday or Sunday, I'd go out there on Route 66 and cook steaks.

Then there were the Christmas cards. One day, only a few weeks after I had arrived at Los Alamos, Arlene says, "I thought it would be nice to send Christmas cards to everybody. Would you like to see the ones I picked out?"

They were nice cards, all right, but inside they said Merry Christmas, from Rich & Putsy. "I can't send these to Fermi and Bethe," I protested. "I hardly even know them!"

"What do *you* care what other people think?"—naturally. So we sent them.

Next year comes around, and by this time I know Fermi. I know Bethe. I've been over at their houses. I've taken care of their kids. We're all very friendly.

Somewhere along the line, Arlene says to me, in a very formal tone, "You haven't asked me about our Christmas cards this year, Richard . . ."

FEAR goes through me. "Uh, well, let's see the cards."

The cards say Merry Christmas and a Happy New Year, from Richard and Arlene Feynman. "Well, that's fine," I say. "They're very nice. They'll go fine for everybody."

"Oh, no," she says. "They won't do for Fermi and Bethe and all those other famous people." Sure enough, she's got another box of cards.

She pulls one out. It says the usual stuff, and then: From Dr. & Mrs. R. P. Feynman.

So I had to send them those.

"What's this formal stuff, Dick?" they laughed. They were happy that she was having such a good time out of it, and that I had no control over it.

Arlene didn't spend all of her time inventing games. She had sent away for a book called *Sound and Symbol in Chinese.* It was a lovely book—I still have it—with about fifty symbols done in beautiful calligraphy, with explanations like "Trouble: three women in a house." She had the right paper, brushes, and ink, and was practicing calligraphy. She had also bought a Chinese dictionary, to get a lot of other symbols.

One time when I came to visit her, Arlene was practicing these things. She says to herself, "No. That one's wrong."

So I, the "great scientist," say, "What do you mean, 'wrong'? It's only a human convention. There's no law of nature which says how they're supposed to look; you can draw them any way you want."

"I mean, artistically it's wrong. It's a question of balance, of how it feels."

"But one way is just as good as another," I protest.

"Here," she says, and she hands me the brush. "Make one yourself."

So I made one, and I said, "Wait a minute. Let me make another one—it's too blobby." (I couldn't say it was wrong, after all.)

"How do you know how blobby it's supposed to be?" she says.

I learned what she meant. There's a particular way you have to make the stroke for it to look good. An aesthetic thing has a certain set, a certain character, which I can't define. Because it couldn't be defined made me think there was nothing to it. But I learned from that experience that there *is* something to it—and it's a fascination I've had for art ever since.

Just at this moment, my sister sends me a postcard from Oberlin, where she's going to college. It's written in pencil, with small symbols—it's in Chinese.

Joan is nine years younger than I am, and studied physics, too. Having me as her older brother was tough on her. She was always looking for something I couldn't do, and was secretly taking Chinese.

Well, I didn't know any Chinese, but one thing I'm

good at is spending an infinite amount of time solving a puzzle. The next weekend I took the card with me to Albuquerque. Arlene showed me how to look up the symbols. You have to start in the back of the dictionary with the right category and count the number of strokes. Then you go into the main part of the dictionary. It turns out each symbol has several possible meanings, and you have to put several symbols together before you can understand it.

With great patience I worked everything out. Joan was saying things like, "I had a good time today." There was only one sentence I couldn't figure out. It said, "Yesterday we celebrated mountain-forming day"—obviously an error. (It turned out they did have some crazy thing called "Mountain-forming Day" at Oberlin, and I had translated it right!)

So it was trivial things like you'd expect to have on a postcard, but I knew from the situation that Joan was trying to floor me by sending me Chinese.

I looked back and forth through the art book and picked out four symbols which would go well together. Then I practiced each one, over and over. I had a big pad of paper, and I would make fifty of each one, until I got it just right.

When I had accidentally made one good example of each symbol, I saved them. Arlene approved, and we glued the four of them end to end, one on top of the other. Then we put a little piece of wood on each end, so you could hang it up on the wall. I took a picture of my masterpiece with Nick Metropolis's camera, rolled up the scroll, put it in a tube, and sent it to Joan.

So she gets it. She unrolls it, and she can't read it. It looks to her as if I simply made four characters, one right after the other, on the scroll. She takes it to her teacher.

The first thing he says is, "This is written rather well! Did you do this?"

"Uh, no. What does it say?"

"Elder brother also speaks."

I'm a real bastard—I would never let my little sister score one on me.

When Arlene's condition became much weaker, her father came out from New York to visit her. It was difficult and expensive to travel that far during the war, but he knew the end was near. One day he telephoned me at Los Alamos. "You'd better come down here right away," he said.

I had arranged ahead of time with a friend of mine at Los Alamos, Klaus Fuchs, to borrow his car in case of an emergency, so I could get to Albuquerque quickly. I picked up a couple of hitchhikers to help me in case something happened on the way.

Sure enough, as we were driving into Santa Fe, we got a flat tire. The hitchhikers helped me change the tire. Then on the other side of Santa Fe, the spare tire went flat, but there was a gas station nearby. I remember waiting patiently for the gas station man to take care of some other car, when the two hitchhikers, knowing the situation, went over and explained to the man what it was. He fixed the flat right away. We decided not to get the spare tire fixed, because repairing it would have taken even more time.

We started out again towards Albuquerque, and I felt foolish that I hadn't thought to say anything to the gas station man when time was so precious. About thirty miles from Albuquerque, we got another flat! We had to abandon the car, and we hitchhiked the rest of the way. I called up a towing company and told them the situation.

I met Arlene's father at the hospital. He had been there for a few days. "I can't take it any more," he said. "I have to go home." He was so unhappy, he just left.

When I finally saw Arlene, she was very weak, and a bit fogged out. She didn't seem to know what was happening. She stared straight ahead most of the time, looking around a little bit from time to time, and was trying to breathe. Every once in a while her breathing would stop—and she would sort of swallow—and then it would start again. It kept going like this for a few hours.

I took a little walk outside for a while. I was surprised that I wasn't feeling what I thought people were supposed to feel under the circumstances. Maybe I was fooling myself. I wasn't delighted, but I didn't feel terribly upset, perhaps because we had known for a long time that it was going to happen.

It's hard to explain. If a Martian (who, we'll imagine, never dies except by accident) came to Earth and saw this peculiar race of creatures—these humans who live about seventy or eighty years, knowing that death is going to come—it would look to him like a terrible problem of psychology to live under those circumstances, knowing that life is only temporary. Well, we humans somehow figure out how to live despite this problem: we laugh, we joke, we live.

The only difference for me and Arlene was, instead of fifty years, it was five years. It was only a quantitative difference—the psychological problem was just the same. The only way it would have become any different is if we had said to ourselves, "But those other people have it better, because they might live fifty years." But that's crazy. Why make yourself miserable saying things like, "Why do we have such bad luck? What has God done to us? What have we done to deserve this?"—all of which, if you understand reality and take it completely into your heart, are irrelevant and unsolvable. They are just things that nobody can know. Your situation is just an accident of life.

We had a hell of a good time together.

I came back into her room. I kept imagining all the things that were going on physiologically: the lungs aren't getting enough air into the blood, which makes the brain fogged out and the heart weaker, which makes the breathing even more difficult. I kept expecting some sort of avalanching effect, with everything caving in together in a dramatic collapse. But it didn't appear that way at all: she just slowly got more foggy, and her breathing gradually became less and less, until there was no more breath—but just before that, there was a very small one.

The nurse on her rounds came in and confirmed that Arlene was dead, and went out—I wanted to be alone for a moment. I sat there for a while, and then went over to kiss her one last time.

I was very surprised to discover that her hair smelled exactly the same. Of course, after I stopped and thought about it, there was no reason why hair should smell different in such a short time. But to me it was a kind of a shock, because in my mind, something enormous had just happened—and yet nothing had happened.

The next day, I went to the mortuary. The guy hands me some rings he's taken from her body. "Would you like to see your wife one last time?" he asks.

"What kind of a—no, I don't want to see her, no!" I said. "I just saw her!"

"Yes, but she's been all fixed up," he says.

This mortuary stuff was completely foreign to me. Fixing up a body when there's nothing there? I didn't want to look at Arlene again; that would have made me more upset.

I called the towing company and got the car, and packed Arlene's stuff in the back. I picked up a hitchhiker, and started out of Albuquerque.

It wasn't more than five miles before . . . BANG! Another flat tire. I started to curse.

The hitchhiker looked at me like I was mentally unbalanced. "It's just a tire, isn't it?" he says.

"Yeah, it's just a tire—and another tire, and again another tire, and another tire!"

We put the spare tire on, and went very slowly, all the way back to Los Alamos, without getting the other tire repaired.

I didn't know how I was going to face all my friends at
Los Alamos. I didn't want people with long faces talking to
me about the death of Arlene. Somebody asked me what
happened.

"She's dead. And how's the program going?" I said.

They caught on right away that I didn't want to moon
over it. Only one guy expressed his sympathy, and it turned
out he had been out of town when I came back to Los
Alamos.

One night I had a dream, and Arlene came into it.
Right away, I said to her, "No, no, you can't be in this
dream. You're not alive!"

Then later, I had another dream with Arlene in it. I
started in again, saying, "You can't be in this dream!"

"No, no," she says. "I fooled you. I was tired of you,
so I cooked up this ruse so I could go my own way. But now
I like you again, so I've come back." My mind was really
working against itself. It had to be explained, even in a
goddamn *dream,* why it was possible that she was still there!

I must have done something to myself, psychologically. I didn't cry until about a month later, when I was
walking past a department store in Oak Ridge and noticed
a pretty dress in the window. I thought, "Arlene would like
that," and then it hit me.

It's as Simple as One, Two, Three...

WHEN I was a kid growing up in Far Rockaway, I had a friend named Bernie Walker. We both had "labs" at home, and we would do various "experiments." One time, we were discussing something—we must have been eleven or twelve at the time—and I said, "But thinking is nothing but talking to yourself inside."

"Oh yeah?" Bernie said. "Do you know the crazy shape of the crankshaft in a car?"

"Yeah, what of it?"

"Good. Now, tell me: how did you describe it when you were talking to yourself?"

So I learned from Bernie that thoughts can be visual as well as verbal.

Later on, in college, I became interested in dreams. I wondered how things could look so real, just as if light were hitting the retina of the eye, while the eyes are closed: are the nerve cells on the retina actually being stimulated in some other way—by the brain itself, perhaps—or does the brain have a "judgment department" that gets slopped up during dreaming? I never got satisfactory answers to such questions from psychology, even though I became very interested in how the brain works. Instead, there was all this business about interpreting dreams, and so on.

When I was in graduate school at Princeton a kind of dumb psychology

paper came out that stirred up a lot of discussion. The
author had decided that the thing controlling the "time
sense" in the brain is a chemical reaction involving iron. I
thought to myself, "Now, how the hell could he figure
that?"

Well, the way he did it was, his wife had a chronic fever
which went up and down a lot. Somehow he got the idea
to test her sense of time. He had her count seconds to
herself (without looking at a clock), and checked how long
it took her to count up to 60. He had her counting—the
poor woman—all during the day: when her fever went up,
he found she counted quicker; when her fever went down,
she counted slower. Therefore, he thought, the thing that
governed the "time sense" in the brain must be running
faster when she's got fever than when she hasn't got fever.

Being a very "scientific" guy, the psychologist knew
that the rate of a chemical reaction varies with the surrounding temperature by a certain formula that depends
on the energy of the reaction. He measured the differences
in speed of his wife's counting, and determined how much
the temperature changed the speed. Then he tried to find
a chemical reaction whose rates varied with temperature in
the same amounts as his wife's counting did. He found that
iron reactions fit the pattern best. So he deduced that his
wife's sense of time was governed by a chemical reaction in
her body involving iron.

Well, it all seemed like a lot of baloney to me—there
were so many things that could go wrong in his long chain
of reasoning. But it *was* an interesting question: what *does*
determine the "time sense"? When you're trying to count
at an even rate, what does that rate depend on? And what
could you do to yourself to change it?

I decided to investigate. I started by counting seconds—without looking at a clock, of course—up to 60 in a
slow, steady rhythm: 1, 2, 3, 4, 5. . . . When I got to 60, only
48 seconds had gone by, but that didn't bother me: the

problem was not to count for exactly one minute, but to count at a standard rate. The next time I counted to 60, 49 seconds had passed. The next time, 48. Then 47, 48, 49, 48, 48. . . . So I found I could count at a pretty standard rate.

Now, if I just sat there, without counting, and waited until I thought a minute had gone by, it was very irregular—complete variations. So I found it's very poor to estimate a minute by sheer guessing. But by counting, I could get very accurate.

Now that I knew I could count at a standard rate, the next question was—what affects the rate?

Maybe it has something to do with the heart rate. So I began to run up and down the stairs, up and down, to get my heart beating fast. Then I'd run into my room, throw myself down on the bed, and count up to 60.

I also tried running up and down the stairs and counting to myself *while* I was running up and down.

The other guys saw me running up and down the stairs, and laughed. "What are you doing?"

I couldn't answer them—which made me realize I couldn't talk while I was counting to myself—and kept right on running up and down the stairs, looking like an idiot.

(The guys at the graduate college were used to me looking like an idiot. On another occasion, for example, a guy came into my room—I had forgotten to lock the door during the "experiment"—and found me in a chair wearing my heavy sheepskin coat, leaning out of the wide-open window in the dead of winter, holding a pot in one hand and stirring with the other. "Don't bother me! Don't bother me!" I said. I was stirring Jell-O and watching it closely: I had gotten curious as to whether Jell-O would coagulate in the cold if you kept it moving all the time.)

Anyway, after trying every combination of running up and down the stairs and lying on the bed, surprise! The heart rate had no effect. And since I got very hot running up and down the stairs, I figured temperature had nothing to do with it either (although I must have known that your temperature doesn't really go up when you exercise). In fact, I couldn't find anything that affected my rate of counting.

Running up and down stairs got pretty boring, so I started counting while I did things I had to do anyway. For instance, when I put out the laundry, I had to fill out a form saying how many shirts I had, how many pants, and so on. I found I could write down "3" in front of "pants" or "4" in front of "shirts," but I couldn't count my socks. There were too many of them: I'm already using my "counting machine"—36, 37, 38—and here are all these socks in front of me—39, 40, 41. . . . How do I count the socks?

I found I could arrange them in geometrical patterns—like a square, for example: a pair of socks in this corner, a pair in that one; a pair over here, and a pair over there—eight socks.

I continued this game of counting by patterns, and found I could count the lines in a newspaper article by grouping the lines into patterns of 3, 3, 3, and 1 to get 10; then 3 of those patterns, 3 of those patterns, 3 of those patterns, and 1 of those patterns made 100. I went right down the newspaper like that. After I had finished counting up to 60, I knew where I was in the patterns and could say, "I'm up to 60, and there are 113 lines." I found that I could even *read* the articles while I counted to 60, and it didn't affect the rate! In fact, I could do anything while counting to myself—except talk out loud, of course.

What about typing—copying words out of a book? I found that I could do that, too, but here my time was affected. I was excited: finally, I've found something that appears to affect my counting rate! I investigated it more.

I would go along, typing the simple words rather fast, counting to myself 19, 20, 21, typing along, counting 27, 28, 29, typing along, until—What the hell is that word?—

Oh, yeah—and then continue counting 30, 31, 32, and so on. When I'd get to 60, I'd be late.

After some introspection and further observation, I realized what must have happened: I would interrupt my counting when I got to a difficult word that "needed more brains," so to speak. My counting rate wasn't slowing down; rather, the counting itself was being held up temporarily from time to time. Counting to 60 had become so automatic that I didn't even notice the interruptions at first.

The next morning, over breakfast, I reported the results of all these experiments to the other guys at the table. I told them all the things I could do while counting to myself, and said the only thing I absolutely could not do while counting to myself was talk.

One of the guys, a fella named John Tukey, said, "I don't believe you can read, and I don't see why you can't talk. I'll bet you I can talk while counting to myself, and I'll bet you you can't read."

So I gave a demonstration: they gave me a book and I read it for a while, counting to myself. When I reached 60 I said, "Now!"—48 seconds, my regular time. Then I told them what I had read.

Tukey was amazed. After we checked him a few times to see what his regular time was, he started talking: "Mary had a little lamb; I can say anything I want to, it doesn't make any difference; I don't know what's bothering you"— blah, blah, blah, and finally, "Okay!" He hit his time right on the nose! I couldn't believe it!

We talked about it a while, and we discovered something. It turned out that Tukey was counting in a different way: he was visualizing a tape with numbers on it going by. He would say, "Mary had a little lamb," and he would *watch* it! Well, now it was clear: he's "looking" at his tape going by, so he can't read, and I'm "talking" to myself when I'm counting, so I can't speak!

After that discovery, I tried to figure out a way of

reading out loud while counting—something neither of us could do. I figured I'd have to use a part of my brain that wouldn't interfere with the seeing or speaking departments, so I decided to use my fingers, since that involved the sense of touch.

I soon succeeded in counting with my fingers and reading out loud. But I wanted the whole process to be mental, and not rely on any physical activity. So I tried to imagine the feeling of my fingers moving while I was reading out loud.

I never succeeded. I figured that was because I hadn't practiced enough, but it might be impossible: I've never met anybody who can do it.

By that experience Tukey and I discovered that what goes on in different people's heads when they *think* they're doing the same thing—something as simple as *counting*—is different for different people. And we discovered that you can externally and objectively test how the brain works: you don't have to ask a person how he counts and rely on his own observations of himself; instead, you observe what he can and can't do while he counts. The test is absolute. There's no way to beat it; no way to fake it.

It's natural to explain an idea in terms of what you already have in your head. Concepts are piled on top of each other: this idea is taught in terms of that idea, and that idea is taught in terms of another idea, which comes from counting, which can be so different for different people!

I often think about that, especially when I'm teaching some esoteric technique such as integrating Bessel functions. When I see equations, I see the letters in colors—I don't know why. As I'm talking, I see vague pictures of Bessel functions from Jahnke and Emde's book, with light-tan j's, slightly violet-bluish n's, and dark brown x's flying around. And I wonder what the hell it must look like to the students.

Getting Ahead

ONE TIME, back in the fifties, when I was returning from Brazil by boat, we stopped off in Trinidad for a day, so I decided to see the main city, Port of Spain. In those days, when I visited a city I was most interested in seeing the poorest sections—to see how life works at the bottom end.

I spent some time off in the hills, in the Negro section of town, wandering around on foot. On the way back a taxi stopped and the driver said, "Hey, mon! You want to see the city? It only cost five biwi."

I said, "Okay," and got in the taxi.

The driver started right off to go up and see some palace, saying, "I'll show you all the fancy places."

I said, "No, thank you; that's similar in every city. I want to see the bottom part of the city, where the poor people live. I've already seen the hills up there."

"Oh!" he said, impressed. "I'll be glad to show you around. And I have a question for you when we're through, so I want you to look at everything carefully."

So he took me to an East Indian neighborhood—it must have been some housing project—and he stopped in front of a house made of concrete blocks. There was practically nothing inside. A man was sitting on the front steps. "You see that man?" he said. "He has a son studyin' medicine in Maryland."

Then he picked up someone from the neighborhood so I could better see what they were like. It was a woman whose teeth had a lot of decay.

Further along we stopped and he introduced me to two women he admired. "They got enough money together to buy a sewing machine, and now they do sewing and tailoring work for people in the neighborhood," he said, proudly. When he introduced me to them, he said, "This man is a professor, and what's interesting is, he wants to see our neighborhoods."

We saw many things, and finally the taxi driver said to me, "Now, Professor, here is my question: you see the Indian people are just as poor, and sometimes even poorer than the Negro people, but they're getting somewhere, somehow—this man has sent his son to college; those women are building up a sewing business. But my people aren't getting anywhere. Why is that?"

I told him, of course, that I didn't know—which is my answer to almost every question—but he wouldn't accept that, coming from a professor. I tried to guess at something which I thought was possible. I said, "There's a long tradition behind life in India that comes from a religion and philosophy that is thousands of years old. And although these people are not in India, they still pass on those traditions about what's important in life—trying to build for the future and supporting their children in the effort—which have come down to them for centuries."

I continued, "I think that your people have unfortunately not had a chance to develop such a long tradition, or if they did, they lost it through conquest and slavery." I don't know if it's true, but it was my best guess.

The taxi driver felt that it was a good observation, and said he was planning to build for the future, too: he had some money on the horses, and if he won, he would buy his own taxicab, and *really* do well.

I felt very sorry. I told him that betting on the horses was a bad idea, but he insisted it was the only way he could do it. He had such good intentions, but his method was going to be luck.

I wasn't going to go on philosophizing, so he took me to a place where there was a steel band playing some great calypso music, and I had an enjoyable afternoon.

ONE TIME, when I was in Geneva, Switzerland, for a Physical Society meeting, I was walking around and happened to go past the United Nations buildings. I thought to myself, "Gee! I think I'll go in and look around." I wasn't particularly dressed for it—I was wearing dirty pants and an old coat—but it turned out there were tours you could go on where some guy would show you around.

The tour was quite interesting, but the most striking part was the great big auditorium. You know how everything is overdone for these big international characters, so what would ordinarily be a stage or a dais was in several layers: you have to climb up whole sequences of steps to this great, big, monstrous wooden thing that you stand behind, with a big screen in back of you. In front of you are the seats. The carpets are elegant, and the big doors with brass handles at the back are beautiful. On each side of the great auditorium, up above, are windowed booths for the translators of different languages to work in. It's a fantastic place, and I kept thinking to myself, "Gee! How it must be to give a talk in a place like this!"

Right after that, we were walking along the corridor just outside the auditorium when the guide pointed through the window and said, "You see those buildings over there that are under construction? They'll be used

Hotel City

for the first time at the Atoms for Peace Conference, in about six weeks."

I suddenly remembered that Murray Gell-Mann and I were supposed to give talks at that conference on the present situation of high-energy physics. My talk was set for the plenary session, so I asked the guide, "Sir, where would the talks for the *plenary* session of that conference be?"

"Back in that room that we just came through."

"Oh!" I said in delight. "Then I'm gonna give a speech in that room!"

The guide looked down at my dirty pants and my sloppy shirt. I realized how dumb that remark must have sounded to him, but it was genuine surprise and delight on my part.

We went along a little bit farther, and the guide said, "This is a lounge for the various delegates, where they often hold informal discussions." There were some small, square windows in the doors to the lounge that you could look through, so people looked in. There were a few men sitting there talking.

I looked through the windows and saw Igor Tamm, a physicist from Russia that I know. "Oh!" I said. "I know that guy!" and I started through the door.

The guide screamed, "No, no! Don't go in there!" By this time he was *sure* he had a maniac on his hands, but he couldn't chase me because he wasn't allowed to go through the door himself!

Tamm's face lit up when he recognized me, and we talked a little bit. The guide was relieved and continued the tour without me, and I had to run to catch up.

At the Physical Society meeting my good friend Bob Bacher said to me, "Listen: it's going to be hard to get a room when that Atoms for Peace Conference is going on. Why don't you have the State Department arrange a room for you, if you haven't already made a reservation?"

"Naw!" I said. "I'm not gonna have the State Department do a damn thing for me! I'll do it myself."

When I returned to my hotel I told them that I would be leaving in a week, but I'd be coming back at the end of summer: "Could I make a reservation now for that time?"

"Certainly! When will you be returning?"

"The second week in September . . ."

"Oh, we're terribly sorry, Professor Feynman; we are already completely booked for that time."

So I wandered off, from one hotel to another, and found they were all booked solid, six weeks ahead of time!

Then I remembered a trick I used once when I was with a physicist friend of mine, a quiet and dignified English fellow.

We were going across the United States by car, and when we got just beyond Tulsa, Oklahoma, there were supposed to be big floods up ahead. We came into this little town and we saw cars parked everywhere, with people and families in them, trying to sleep. He says, "We had better stop here. It's clear we can go no further."

"Aw, come on!" I say. "How do you know? Let's see if we can *do* it: maybe by the time we get there, the water will be down.

"We shouldn't waste time," he replies. "Perhaps we can find a room in a hotel if we look for it now."

"Aw, don't worry about it!" I say. "Let's go!"

We drive out of town about ten or twelve miles and come to an arroyo. Yes, even for me, there's too much water. There's no question: we aren't going to try to get through *that*.

We turn around: my friend's muttering about how we'll have no chance of finding a room in a hotel now, and I tell him not to worry.

Back in town, it's absolutely *blocked* with people sleeping in their cars, obviously because there are no more rooms. All the hotels must be packed. I see a small sign over a door: it says "HOTEL." It was the kind of hotel I was familiar with in Albuquerque, when I would wander around town looking at things, waiting to see my wife at the hospital: you have to go up a flight of stairs and the office is on the first landing.

We go up the stairs to the office and I say to the manager, "We'd like a room."

"Certainly, sir. We have one with two beds on the third floor."

My friend is amazed: The town is packed with people sleeping in cars, and here's a hotel that has room!

We go up to our room, and gradually it becomes clear to him: there's no door on the room, only a hanging cloth in the doorway. The room was fairly clean, it had a sink; it wasn't so bad. We get ready for bed.

He says, "I've got to pee."

"The bathroom is down the hall."

We hear girls giggling and walking back and forth in the hall outside, and he's nervous. He doesn't want to go out there.

"That's all right; just pee in the sink," I say.

"But that's unsanitary."

"Naw, it's okay; you just turn the water on."

"I can't pee in the sink," he says.

We're both tired, so we lie down. It's so hot that we don't use any covers, and my friend can't get to sleep because of the noises in the place. I kind of fall asleep a little bit.

A little later I hear a creaking of the floor nearby, and I open one eye slightly. There he is, in the dark, quietly stepping over to the sink.

Anyway, I knew a little hotel in Geneva called the Hotel City, which was one of those places with just a doorway on the street and a flight of stairs leading up to the office. There were usually some rooms available, and nobody made reservations.

I went up the stairs to the office and told the desk clerk that I'd be back in Geneva in six weeks, and I'd like to stay in their hotel: "Could I make a reservation?"

"Certainly, sir. Of course!"

The clerk wrote my name on a piece of paper—they hadn't any book to write reservations in—and I remember the clerk trying to find a hook to put the paper on, to remember. So I had my "reservation," and everything was fine.

I came back to Geneva six weeks later, went to the Hotel City, and they *did* have the room ready for me; it was on the top floor. Although the place was cheap, it was clean. (It's Switzerland; it was *clean!*) There were a few holes in the bedspread, but it was a clean bedspread. In the morning they served a European breakfast in my room; they were rather delighted to have this guest who had made a reservation six weeks in advance.

Then I went over to the U.N. for the first day of the Atoms for Peace Conference. There was quite a line at the reception desk, where everyone was checking in: a woman was taking down everybody's address and phone number so they could be reached in case there were any messages.

"Where are you staying, Professor Feynman?" she asks.

"At the Hotel City."

"Oh, you must mean the Hotel Cité."

"No, it's called 'City': C-I-T-Y." (Why not? We would call it "Cité" here in America, so they called it "City" in Geneva, because it sounded foreign.)

"But it isn't on our list of hotels. Are you *sure* it's 'City'?"

"Look in the telephone book for the number. You'll find it."

"Oh!" she said, after checking the phone book. "My list is incomplete! Some people are still looking for a room, so perhaps I can recommend the Hotel City to them."

She must have got the word about the Hotel City from

someone, because nobody else from the conference ended up staying there. Once in a while the people at the Hotel City would receive telephone calls for me from the U.N., and would *run up* the two flights of stairs from the office to tell me, with some awe and excitement, to come down and answer the phone.

There's an amusing scene I remember from the Hotel City. One night I was looking through my window out into the courtyard. Something, in a building across the courtyard, caught the corner of my eye: it looked like an upside-down bowl on the windowsill. I thought it had moved, so I watched it for a while, but it didn't move any. Then, after a bit, it moved a little to one side. I couldn't figure out what this thing was.

After a while I figured it out: it was a man with a pair of binoculars that he had against the windowsill for support, looking across the courtyard to the floor below me!

There's another scene at the Hotel City which I'll always remember, that I'd love to be able to paint: I was returning one night from the conference and opened the door at the bottom of the stairway. There was the proprietor, standing there, trying to look nonchalant with a cigar in one hand while he pushed something up the stairs with the other. Farther up, the woman who brought me breakfast was pulling on this same heavy object with both hands. And at the top of the stairs, at the landing, there *she* was, with her fake furs on, bust sticking out, hand on her hip, *imperiously waiting.* Her customer was a bit drunk, and was not very capable of walking up the steps. I don't know whether the proprietor knew that *I* knew what this was all about; I just walked past everything. He was ashamed of his hotel, but, of course, to me, it was delightful.

ONE DAY I got a long-distance telephone call from an old friend in Los Alamos. She says in a very serious voice, "Richard, I have some sad news for you. Herman died."

I'm always feeling uncomfortable that I don't remember names and then I feel bad that I don't pay enough attention to people. So I said, "Oh?"—trying to be quiet and serious so I could get more information, but thinking to myself, "Who the hell is Herman?"

She says, "Herman and his mother were both killed in an automobile accident near Los Angeles. Since that is where his mother is from, the funeral will be held in Los Angeles at the Rose Hills Mortuary on May 3rd at three o'clock." Then she says, "Herman would have liked it very, very much to know that you would be one of his pallbearers."

I still can't remember him. I say, "Of course I'd be happy to do that." (At least this way I'll find out who Herman is.)

Then I get an idea: I call up the mortuary. "You're having a funeral on May 3rd at three o'clock . . ."

"Which funeral do you mean: the Goldschmidt funeral, or the Parnell funeral?"

"Well, uh, I don't know." It still doesn't click for me; I don't think it's either one of them. Finally, I say, "It

Who the Hell Is Herman?

might be a double funeral. His mother also died."

"Oh, yes. Then it's the Goldschmidt funeral."

"Herman Goldschmidt?"

"That's right; Herman Goldschmidt and Mrs. Goldschmidt."

Okay. It's Herman Goldschmidt. But I still can't remember a Herman Goldschmidt. I haven't any idea what it is I've forgotten; from the way she talked, my friend was sure that Herman and I knew each other well.

The last chance I have is to go to the funeral and look into the casket.

I go to the funeral, and the woman who had arranged everything comes over, dressed in black, and says in a sorrowful voice, "I'm *so* glad you're here. Herman would be so happy if he knew"—all this serious stuff. Everybody's got long faces about Herman, but I still don't know who Herman is—though I'm sure that if I knew, I'd feel very sorry that he was dead!

The funeral proceeded, and when it came time for everybody to file past the caskets, I went up. I looked into the first casket, and there was Herman's mother. I looked into the second casket, and there was Herman—and I swear to you, I'd never seen him before in my life!

It came time to carry the casket out, and I took my place among the pallbearers. I very carefully laid Herman to rest in his grave, because I knew he would have appreciated it. But I haven't any idea, to this day, who Herman was.

Many years later I finally got up enough courage to bring it up to my friend. "You know that funeral I went to, about ten years ago, for Howard . . ."

"You mean Herman."

"Oh yeah—Herman. You know, I didn't know who Herman was. I didn't even recognize him in the casket."

"But Richard, you knew each other in Los Alamos just

after the war. You were both good friends of mine, and we had many conversations together."

"I still can't remember him."

A few days later she called and told me what might have happened: maybe she had met Herman just after I had left Los Alamos—and therefore got the timing mixed up somehow—but because she was such good friends with each of us, she thought we must have known each other. So she was the one who had made the mistake, not me (which is usually the case). Or was she just being polite?

Feynman Sexist Pig!

A FEW YEARS after I gave some lectures for the freshmen at Caltech (which were published as the *Feynman Lectures on Physics*), I received a long letter from a feminist group. I was accused of being anti-woman because of two stories: the first was a discussion of the subtleties of velocity, and involved a woman driver being stopped by a cop. There's a discussion about how fast she was going, and I had her raise valid objections to the cop's definitions of velocity. The letter said I was making the woman look stupid.

The other story they objected to was told by the great astronomer Arthur Eddington, who had just figured out that the stars get their power from burning hydrogen in a nuclear reaction producing helium. He recounted how, on the night after his discovery, he was sitting on a bench with his girlfriend. She said, "Look how pretty the stars shine!" To which he replied, "Yes, and right now, I'm the only man in the world who knows *how* they shine." He was describing a kind of wonderful loneliness you have when you make a discovery.

The letter claimed that I was saying a woman is incapable of understanding nuclear reactions.

I figured there was no point in trying to answer their accusations in detail, so I wrote a short letter back to them: "Don't bug me, man!"

   Needless to say, that didn't work too well. Another letter came: "Your response to our letter of September 29th is unsatisfactory . . ."—blah, blah, blah. This letter warned that if I didn't get the publisher to revise the things they objected to, there would be trouble.

   I ignored the letter and forgot about it.

   A year or so later, the American Association of Physics Teachers awarded me a prize for writing those books, and asked me to speak at their meeting in San Francisco. My sister, Joan, lived in Palo Alto—an hour's drive away—so I stayed with her the night before and we went to the meeting together.

   As we approached the lecture hall, we found people standing there giving out handbills to everybody going in. We each took one, and glanced at it. At the top it said, "A PROTEST." Then it showed excerpts from the letters they sent me, and my response (in full). It concluded in large letters: "FEYNMAN SEXIST PIG!"

   Joan stopped suddenly and rushed back: "These are interesting," she said to the protester. "I'd like some more of them!"

   When she caught up with me, she said, "Gee whiz, Richard; what did you do?"

   I told her what had happened as we walked into the hall.

   At the front of the hall, near the stage, were two prominent women in the American Association of Physics Teachers. One was in charge of women's affairs for the organization, and the other was Fay Ajzenberg, a professor of physics I knew, from Pennsylvania. They saw me coming down towards the stage accompanied by this woman with a fistful of handbills, talking to me. Fay walked up to her and said, "Do you realize that Professor Feynman has a *sister* that he encouraged to go into physics, and that *she* has a Ph.D. in physics?"

   "Of course I do," said Joan. "I'm that sister!"

Fay and her associate explained to me that the protesters were a group—led by a man, ironically—who were always disrupting meetings in Berkeley. "We'll sit on either side of you to show our solidarity, and just before you speak, I'll get up and say something to quiet the protesters," Fay said.

Because there was another talk before mine, I had time to think of something to say. I thanked Fay, but declined her offer.

As soon as I got up to speak, half a dozen protesters marched down to the front of the lecture hall and paraded right below the stage, holding their picket signs high, chanting, "Feynman sexist pig! Feynman sexist pig!"

I began my talk by telling the protesters, "I'm sorry that my short answer to your letter brought you here unnecessarily. There are more serious places to direct one's attention towards improving the status of women in physics than these relatively trivial mistakes—if that's what you want to call them—in a textbook. But perhaps, after all, it's good that you came. For women do indeed suffer from prejudice and discrimination in physics, and your presence here today serves to remind us of these difficulties and the need to remedy them."

The protesters looked at one another. Their picket signs began to come slowly down, like sails in a dying wind.

I continued: "Even though the American Association Physics Teachers has given me an award for teaching, I must confess I don't know how to teach. Therefore, I have nothing to say about teaching. Instead, I would like to talk about something that will be especially interesting to the women in the audience: I would like to talk about the structure of the proton."

The protesters put their picket signs down and walked off. My hosts told me later that the man and his group of protesters had never been defeated so easily.

(Recently I discovered a transcript of my speech, and

what I said at the beginning doesn't seem anywhere near as dramatic as the way I remember it. What I remember saying is much more wonderful than what I actually said!)

After my talk, some of the protesters came up to press me about the woman-driver story. "Why did it have to be a woman driver?" they said. "You are implying that all women are bad drivers."

"But the woman makes the cop look bad," I said. "Why aren't you concerned about the cop?"

"That's what you expect from cops!" one of the protesters said. "They're all pigs!"

"But you *should* be concerned," I said. "I forgot to say in the story that the cop was a woman!"

I Just Shook His Hand, Can You Believe It?

FOR SOME YEARS now the University of Tokyo has been inviting me to visit Japan. But every time I accepted their invitation, I would happen to get sick and not be able to go.

In the summer of 1986 there was going to be a conference in Tokyo, and the university again invited me to come. Although I love Japan and wanted very much to visit, I felt uncomfortable at the invitation because I had no paper to give. The university said it would be all right for me to give a summary paper, but I said I don't like to do that. But then they said they would be honored if I would be the chairman of one session of the conference—that's all I would have to do. So I finally said okay.

I was lucky this time and didn't get sick.* So Gweneth and I went to Tokyo, and I was chairman of one session.

The chairman is supposed to make sure that the speakers only talk for a certain length of time, in order to leave enough time for the next speaker. The chairman occupies a position of such high honor that there are two co-chairmen to assist him. My co-chairmen said they would take care of introducing the speakers, as well as telling them when it's time to stop.

*Feynman was suffering from abdominal cancer. He had surgery in 1978 and 1981. After he returned from Japan, he had more surgery, in October 1986 and October 1987.

Things went smoothly for most of the session until one speaker—a Japanese man—didn't stop talking when his time was up. I look at the clock and figure it's time he should stop. I look over at the co-chairmen and gesture a little bit.

They come up to me and say, "Don't do anything; we'll take care of it. He's talking about Yukawa.* It's all right."

So I was the honorary chairman of one session, and I felt I didn't even do my job right. And for that, the university paid my way to Japan, they took care of arranging my trip, and they were all very gracious.

One afternoon we were talking to the host who was arranging our trip. He shows us a railroad map, and Gweneth sees a curved line with lots of stops in the middle of the Ise Peninsula—it's not near the water; it's not near anywhere. She puts her finger on the end of the line and says, "We want to go here."

He looks at it, and says, "Oh! You want to go to . . . Iseokitsu?"

She says, "Yes."

"But there's *nothing* in Iseokitsu," he says, looking at me as if my wife is crazy, and hoping I'll bring her back to her senses.

So I say, "Yes, that's right; we want to go to Iseokitsu."

Gweneth hadn't talked to me about it, but I knew what she was thinking: we enjoy traveling to places in the middle of nowhere, places we've never heard of, places which have nothing.

Our host becomes a little bit upset: he's never made a hotel reservation for Iseokitsu; he doesn't even know if there's an inn there.

He gets on the telephone and calls up Iseokitsu for us. In Iseokitsu, it turns out, there are no accommodations.

*Hideki Yukawa. Eminent Japanese physicist; Nobel Prize, 1949.

But there's another town—about seven kilometers beyond the end of the line—that has a Japanese-style inn.

We say, "Fine! That's just what we want—a Japanese-style inn!" They give him the number and he calls.

The man at the inn is very reluctant: "Ours is a very small inn. It's a family-run place."

"That's what they want," our host reassures him.

"Did he say yes?" I ask.

After more discussion, our host says, "He agrees."

But the next morning, our host gets a telephone call from this same inn: last night they had a family conference. They decided they can't handle the situation. They can't take care of foreigners.

I say, "What's the trouble?"

Our host telephones the inn and asks what the problem is. He turns to us and says, "It's the toilet—they don't have a Western-style toilet."

I say, "Tell them that the last time my wife and I went on a trip, we carried a small shovel and toilet paper, and dug holes for ourselves in the dirt. Ask him, 'Shall we bring our shovel?' "

Our host explains this over the telephone, and they say, "It's okay. You can come for one night. You don't need to bring your shovel."

The innkeeper picked us up at the railroad station in Iseokitsu and took us to his inn. There was a beautiful garden outside our room. We noticed a brilliant, emerald-green tree frog climbing a metal frame with horizontal bars (used for hanging out the wet clothes), and a tiny yellow snake in a shrub in front of our *engawa* (veranda). Yes, there was "nothing" in Iseokitsu—but everything was beautiful and interesting to us.

It turned out there was a shrine about a mile away—that's why this little inn was there—so we walked to it. On our way back, it began to rain. A guy passed us in his car, then turned around and came back. "Where are you going?" he asked in Japanese. "To the inn," I said. So he took us there.

When we got back to our room, we discovered that Gweneth had lost a roll of film—perhaps in the man's car. So I got the dictionary out and looked up "film" and "lost," and tried to explain it to the innkeeper. I don't know how he did it, but he found the man who had given us the ride, and in his car we found the film.

The bath was interesting; we had to go through another room to get to it. The bathtub was wooden, and around it were all kinds of little toys—little boats and so on. There was also a towel with Mickey Mouse on it.

The innkeeper and his wife had a little daughter who was two, and a small baby. They dressed their daughter in a kimono and brought her up to our room. Her mother made origami things for her; I made some drawings for her, and we played with her.

A lady across the street gave us a beautiful silk ball that she had made. Everything was friendly; everything was very good.

The next morning we were supposed to leave. We had a reservation at one of the more famous resorts, at a spa somewhere. I looked in the dictionary again; then I came down and showed the innkeeper the receipt for our reservation at the big resort hotel—it was called the Grand View, or something like that. I said, "We don't want stay big hotel tomorrow night; we want stay *here* tomorrow night. We happy here. Please you call them; change this."

He says, "Certainly! Certainly!" I could tell he was pleased by the idea that these foreigners were canceling their reservation in this big, fancy hotel in order to stay in his little inn another night.

After we returned to Tokyo, we went to the University of Kanazawa. Some professors arranged to drive us along the coast of nearby Noto Peninsula. We passed through

several delightful fishing villages, and went to visit a pa-
goda in the middle of the countryside.

Then we visited a shrine with an enclave behind it,
where one could go only by special invitation. The Shinto
priest there was very gracious and invited us into his private
rooms for tea, and he did some calligraphy for us.

After our hosts had taken us a little farther along the
coast, they had to return to Kanazawa. Gweneth and I decided to stay in Togi for two or three days. We stayed in
a Japanese-style hotel, and the lady innkeeper there was
very, very nice to us. She arranged for her brother to take
us by car down the coast to several villages, and then we
came back by bus.

The next morning the innkeeper told us there was
something important happening in town. A new shrine,
replacing an old one, was being dedicated.

When we arrived at the grounds we were invited to sit
on a bench, and were served tea. There were many people
milling around, and eventually a procession came out from
behind the shrine. We were delighted to see the leading
figure was the head priest from the shrine we had visited
a few days before. He was dressed in a big, ceremonial
outfit, and was obviously in charge of everything.

After a little while the ceremony began. We didn't want
to intrude into a religious place, so we stayed back from the
shrine itself. But there were kids running up and down the
steps, playing and making noise, so we figured it wasn't so
formal. We came a little closer and stood on the steps so
we could see inside.

The ceremony was wonderful. There was a ceremonial
cup with branches and leaves on it; there was a group of
girls in special uniforms; there were dancers, and so on. It
was quite elaborate.

We're watching all these performances when all of a
sudden we feel a tap on the shoulder. It's the head priest!
He gestures to us to follow him.

We go around the shrine and enter from the side. The head priest introduces us to the mayor and other dignitaries, and invites us to sit down. A *noh* actor does a dance, and all kinds of other wonderful things go on.

Then there are speeches. First, the mayor gives a speech. Then the head priest gets up to speak. He says, "Unano, utsini kuntana kanao. Untanao uni kanao. Uniyo zoimasu doi zinti Fain-man-san-to unakano kane gozaimas . . ."—and he points to "Fain-man-san" and tells me to say something!

My Japanese is very poor, so I say something in English: "I love Japan," I say. "I am particularly impressed by your tremendous rate of technological change, while at the same time your traditions still mean so much, as you are showing with this shrine dedication." I tried to express the mixture I saw in Japan: change, but without losing respect for traditions.

The head priest says something in Japanese which I do not believe is what I said (although I couldn't really tell), because he had never understood anything I had said to him previously! But he *acted* as if he understood *exactly* what I said, and he "translated" it with complete confidence for everyone. He was much like I am, in this respect.

Anyway, the people politely listened to whatever it was that I said, and then another priest gave a speech. He was a young man, a student of the head priest, dressed in a wonderful outfit with big, wide pant legs and a big, wide hat. He looked so gorgeous, so wonderful.

Then we went to lunch with all the dignitaries, and felt very honored to be included.

After the shrine dedication ceremony was over, Gweneth and I thanked the head priest and left the dining hall to walk around the village for a while. After a bit we found some people pulling a big wagon, with a shrine in it, through the streets. They're all dressed up in outfits with symbols on the back, singing, "Eyo! Eyo!"

We follow the procession, enjoying the festivities, when a policeman with a walkie-talkie comes up to us. He takes off his white glove and puts out his hand. I shake hands with him.

As we leave the policeman and begin to follow the procession again, we hear a loud, high-pitched voice behind us, speaking very rapidly. We turn around and see the policeman clutching his walkie-talkie, speaking into it with great excitement: "O gano fana miyo ganu Fain-man-san iyo kano muri tono muroto kala . . ."—and I could just imagine him telling the person at the other end: "Do you remember that Mr. Fain-man who spoke at the shrine dedication? I just shook his hand, can you believe it?"

The priest must have "translated" something very impressive!

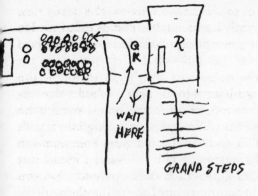

WAIT HERE

GRAND STEPS

Letters, Photos, and Drawings

October 11, 1961
Hotel Amigo, Brussels

Hello, my sweetheart,

Murray and I kept each other awake arguing until we could stand it no longer. We woke up over Greenland, which was even better than last time because we went right over part of it. In London we met other physicists and came to Brussels together. One of them was worried—in his guidebook the Hotel Amigo was not even mentioned. Another had a newer guide— five stars, and rumored to be the best hotel in Europe!

It is very nice indeed. All the furniture is dark red polished wood, in perfect condition; the bathroom is grand, etc. It is really too bad you didn't come to this conference instead of the other one.

At the meeting next day things

started slowly. I was to talk in the afternoon. That is what I did, but I didn't really have enough time. We had to stop at 4 pm because of a reception scheduled for that night. I think my talk was OK though—what I left out was in the written version anyway.

So that evening we went to the palace to meet the king and queen. Taxis waited for us at the hotel—long black ones—and off we went at 5 pm, arriving through the palace gates with a guard on each side, and driving under an arch where men in red coats and white stockings with a black band and gold tassel under each knee opened the doors. More guards at the entrance, in the hallway, along the stairs, and up into a sort of ballroom. These guards, in dark grey Russian-type hats with chin straps, dark coats, white pants, and shiny black leather boots, stand very straight—each holding a sword straight up.

In the "ballroom" we had to wait perhaps 20 minutes. It has inlaid parquet floors, and L in each square (for Leopold—the present king is Baudoin, or something). The gilded walls are 18th century and on the ceiling are pictures of naked women riding chariots among the clouds. Lots of mirrors and gilded chairs with red cushions around the outside edge of the room—just like so many of those palaces we have seen, but this time it's no museum: it's alive, with everything clear and shining, and in perfect condition. Several palace officials were milling around among us. One had a list and told me where to stand but I didn't do it right and was out of place later.

The doors at the end of the hall open. Guards are there with the king and queen; we all enter slowly and are introduced one by one to the king and queen. The king has a young semi-dopey face and a strong handshake; the queen is very pretty. (I think her name is Fabriola—a Spanish countess she was.) We exit into another room on the left where there are lots of chairs arranged like in a theatre, with two in front, also facing forward, for K & Q. A table at the front with six seats is for illustrious scientists—Niels Bohr, J. Perrin (a Frenchman), J. R. Oppenheimer etc.—see drawing.

It turns out the king wants to know what we are doing, so the old boys give a set of six dull lectures—all very solemn—no jokes. I had great difficulty sitting in my seat because I had a very stiff and uncomfortable back from sleeping on the plane.

That done, the K & Q pass through the room where we met them and into a room on right (marked R). (All these rooms are very big, gilded, Victorian, fancy, etc.) In R are many kinds of uniforms: guards at door in red coats, waiters in white coats (to serve drinks and hors d'oeuvres), military khaki and medals, and black coats—undertaker's type (palace officials).

On the way out of L into R, I am last because I walk slowly from stiff back. I find myself talking to a palace official—nice man. He teaches math part time at Louvain University, but his main job is secretary to the queen. He had also tutored the K when K was young and has been in palace work 23 years. Now, at least, I have somebody to talk to.

Some others are talking to K or to Q; everybody is standing up. After a while the professor who is head of the conference (Prof. Bragg) grabs me and says K wants to talk to me. Bragg says, "K, this is Feynman." I pull boner #1 by wanting to shake hands again—apparently wrong: no hand reaches up. After an embarrassed pause K saves day by shaking my hand. K makes polite remarks on how smart we must all be and how hard it must be to think. I answer, making jokes (having been instructed to do so by Bragg, but what does he know?)—apparently error #2. Anyway, strain is relieved when Bragg brings over some other professor—Heisenberg, I think. K forgets F and F slinks off to resume conversation with sec'y of Q.

After considerable time—several orange juices and

many very good hors d'oeuvres later—a military uniform with medals comes over to me and says, "Speak to the queen!" Nothing I should like to do better (pretty girl, but don't worry, she's married). F arrives at scene: Q is sitting at table surrounded by three other occupied chairs—no room for F. There are several low coughs, slight confusion, etc., and lo!—one of the chairs has been reluctantly vacated. Other two chairs contain one lady and one Priest in Full Regalia (who is also a physicist) named LeMaître.

We have quite a conversation (I listen, but hear no coughs, and am not evacuated from seat) for perhaps 15 minutes. Sample:

Q: "It must be very hard work thinking about those difficult problems . . ."

F: "No, we all do it for the fun of it."

Q: "It must be hard to learn to change all your ideas"—(a thing she got from the six lectures).

F: "No, all those guys who gave you those lectures are old fogeys—all that change was in 1926, when I was only eight. So when I learned physics I only had to learn the new ideas. The big problem now is, do we have to change them again?"

Q: "You must feel good, working for peace like that."

F: "No, that never enters my head, whether it is for peace or otherwise. We don't know."

Q: "Things certainly change fast—many things have changed in the last hundred years."

F: "Not in this palace." (I thought it, but controlled myself.) "Yes," and then launched into lecture on what was known in 1861 and what we found out since—adding at end, laughingly, "Can't help giving a lecture, I guess—I'm a professor, you see. Ha, ha."

Q, in desperation, turns to lady on her other side and begins conversation with same.

After a few moments K comes over and whispers something to Q, who stands up—they quietly go out. F returns to sec'y of Q who personally escorts him out of palace past guards, etc.

I'm so terribly sorry you missed it. I don't know when we'll find another king for you to meet.*

I was paged in the hotel this morning just before leaving with the others. I returned to the others and announced, "Gentlemen, that call was from the queen's secretary. I must leave you now." All are awestruck, for it did not go unnoticed that F talked longer and harder to Q than seemed proper. I didn't tell them, however, that it was about a meeting we arranged—he was inviting me to his home to meet his wife and two (of four) of his daughters, and to see his house. I had invited him to visit us in Pasadena when he came to America and this was his response.

His wife and daughters are very nice and his house is positively beautiful. You would have enjoyed that even more than visiting the palace. He planned and built his house in a Belgian style, somewhat after an old farmhouse style, but done just right. He has many old cabinets and tables inside, right beside newer stuff, very well combined. It is much easier to find antiques in Belgium than in Los Angeles as there are so many old farms, etc. The house is slightly bigger than ours and the grounds are much bigger but not yet landscaped, except for a vegetable garden. He has a bench that he made for himself in the garden, hidden under trees, to go and sit on and look at the surrounding countryside. He has a dog—from Washington—that somebody gave to the king and the K gave to him. The dog has a personality somewhat like Kiwi† because I think he is equally loved.

I told the secretary I had a queen in a little castle in Pasadena that I would like him to see, and he said he hoped

*Four years later Richard and Gweneth met the king of Sweden—at the Nobel Prize ceremony.
†The Feynmans' dog.

he would be able to come to America and see us. He would come if the Q ever visits America again.

I am enclosing a picture of his house, and his card, so I don't lose it.

I know you must feel terrible being left out this time— but I'll make it up someday somehow. But don't forget I love you very much and am proud of my family that is and my family that is to be.* The secretary and his wife send their best wishes to you and our future.

I wish you were here, or, next best thing, that I were there. Kiss SNORK† and tell Mom all about my adventures and I will be home sooner than you think.

> Your husband loves you.

> Your husband.

> Grand Hotel
> Warsaw

Dearest Gweneth,

To begin with, I love you.

Also I miss you and the baby‡ and Kiwi, and really wish I were home.

I am now in the restaurant of the Grand Hotel. I was warned by friends that the service is slow, so I went back for pens and paper to work on my talk for tomorrow—but what could be better than to write to my darling instead?

What is Poland like? My strongest impression—and the one which gives me such a surprise—is that it is almost exactly as I pictured it (except for one detail)—not only in

*Gweneth was expecting Carl at the time.
†Kiwi.
‡Carl. This letter was written in 1963.

how it looks, but also in the people, how they feel, what they say and think about the government, etc. Apparently we are well informed in the US and magazines such as Time and Atlas are not so bad. The detail is that I had forgotten how completely destroyed Warsaw was during the war and therefore that, with few exceptions (which are easily identified by the bullet holes all over them), all the buildings are built since the war. In fact it is a rather considerable accomplishment—there are very many new buildings: Warsaw is a big city, all rebuilt.

The genius of builders here is to be able to build old buildings. There are buildings with facings falling off (walls covered with concrete with patches of worn brick showing thru), rusted window bars with streaks of rust running down the building, etc. Further, the architecture is old—decorations sort of 1927 but heavier—nothing interesting to look at (except one building).

The hotel room is very small, with cheap furniture, a very high ceiling (15 feet), old water spots on the walls, plaster showing through where bed rubs wall, etc. It reminds me of an old "Grand Hotel" in New York—faded cotton bedspread covering bumpy bed, etc. But the bathroom fixtures (faucets etc.) are bright and shiny, which confused me: they seem relatively new in this old hotel. I finally found out: the hotel is only three years old—I had forgotten about their ability to build old things. (No attention at all yet from waiter, so I break down and ask a passing one for service. A confused look—he calls another over. Net result: I am told there is no service at my table and am asked to move to another. I make angry noises. The response: I am put at another table, given a menu, and have 15 seconds to make up my mind. I order Sznycel Po Wiedensku—Wiener Schnitzel.)

On the question of whether the room is bugged: I look for covers of old sockets (like the one in the ceiling of the shower). There are five of them, all near the ceiling—15 feet. I need a ladder and decide not to investigate them. But there is a similar large square plate in the lower corner of my room near the telephone. I pull it back a little (one screw is loose). I have rarely seen so many wires—like the back of a radio. What is it? Who knows! I didn't see any microphones; the ends of the wires were taped, like connections or outlets no longer in use. Maybe the microphone is in the tape. Well, I haven't a screwdriver so I don't take the plate off to investigate further. In short, if my room isn't bugged they are wasting a lot of wires.

The Polish people are nice, poor, have at least medium style in (soup arrives!) clothes, etc. There are nice places to dance, with good bands, etc., etc. So Warsaw is not very heavy and dull, as one hears Moscow is. On the other hand, you meet at every turn that kind of dull stupid backwardness characteristic of government—you know, like the fact that change for $20 isn't available when you went to get your card renewed at the US Immigration Office downtown. Example: I lost my pencil, and wanted to buy a new one at the kiosk here. "A pen costs $1.10."

"No, I want a pencil—wooden, with graphite."

"No, only $1.10 pens."

"OK, how many Zlotys is that?"

"You can't buy it in Zlotys, only for $1.10." (Why? Who knows!)

I have to go upstairs for American money. I give $1.25.

Clerk at kiosk cannot give change—must go to cashier of hotel. The bill for my pen is written in quadruplicate: the clerk keeps one, the cashier one, and I get two copies. What shall I do with them? On the back it says I should keep them to avoid paying US customs duties. It is a Papermate pen made in the USA. (The soup dish is removed.)

The real question of government versus private enterprise is argued on too philosophical and abstract a basis. Theoretically, planning may be good. But nobody has ever figured out the cause of government stupidity—and until they do (and find the cure), all ideal plans will fall into quicksand.

I didn't guess right the nature of the palace in which the meetings are held. I imagined an old, forbidding, large room from 16th century or so. Again, I forgot that Poland was so thoroughly destroyed. The palace is brand new: we meet in a round room with white walls, with gilded decorations on the balcony; the ceiling is painted with a blue sky and clouds. (The main course comes. I eat it; it is very good. I order dessert: pastries with pineapple, 125 g. Incidentally, the menu is very precise: the "125 g" is the weight—125 grams. There are things like "filet of herring, 144 g," etc. I haven't seen anybody checking for cheating with a scale; I didn't check if the schnitzel was the claimed 100 grams.)

I am not getting anything out of the meeting. I am learning nothing. Because there are no experiments this field is not an active one, so few of the best men are doing work in it. The result is that there are hosts of dopes here (126) and it is not good for my blood pressure: such inane things are said and seriously discussed that I get into arguments outside the formal sessions (say, at lunch) whenever anyone asks me a question or starts to tell me about his "work." The "work" is always: (1) completely un-understandable, (2) vague and indefinite, (3) something correct that is obvious and self-evident, but worked out by a long and difficult analysis, and presented as an important discovery, or (4) a claim based on the stupidity of the author that some obvious and correct fact, accepted and checked for years, is, in fact, false (these are the worst: no argument will convince the idiot), (5) an attempt to do something probably impossible, but certainly of no utility, which, it is finally revealed at the end, fails (dessert arrives and is eaten), or (6) just plain wrong. There is a great deal of "activity in the field" these days, but this "activity" is mainly in showing that the previous "activity" of somebody else resulted in an error or in nothing useful or in something promising. It is like a lot of worms trying to get out of a bottle by crawling all over each other. It is not that the subject is hard; it is that the good men are occupied elsewhere. Remind me not to come to any more gravity conferences!

I went one evening to the home of one of the Polish professors (young, with a young wife). People are allowed seven square yards per person in apartments, but he and his wife are lucky: they have twenty-one*—for living room, kitchen, bathroom. He was a little nervous with his guests (myself, Professor and Mrs. Wheeler, and another) and seemed apologetic that his apartment was so small. (I ask for the check. All this time the waiter has had two or three active tables, including mine.) But his wife was very relaxed and kissed her siamese cat "Booboosh" just like you do with Kiwi. She did a wonderful job of entertaining—the table for eating had to be taken from the kitchen, a trick requiring the bathroom door to be first removed from its hinges. (There are only four active tables in the whole restaurant now, and four waiters.) Her food was very good and we all enjoyed it.

Oh, I mentioned that one building in Warsaw is interesting to look at. It is the largest building in Poland: the "Palace of Culture and Science," given as a gift by the Soviet Union. It was designed by Soviet architects. Darling, it is unbelievable! I cannot even begin to describe it. It is the craziest monstrosity on land! (The check comes—brought by a different waiter. I await the change.)

This must be the end of my letter. I hope I don't wait too long for the change. I skipped coffee because I thought it would take too long. Even so, see what a long letter I can write while eating Sunday dinner at the Grand Hotel.

*About 200 square feet.

I say again I love you, and wish you were here—or better I were there. Home is good.

(The change has come—it is slightly wrong (by 0.55 Zloty = 15¢) but I let it go.)

Good bye for now.

Richard.

Saturday, June 29(?) 3pm
Royal Olympic Hotel. Poolside.

Dear Gweneth, and Michelle* (and Carl?),

This is my third day in Athens.

I'm writing by the side of the hotel pool with the paper in my lap because the tables are too high and the chairs too low.

The trip was all on time but uncomfortable anyway because the plane from New York to Athens was absolutely full—every seat. I was met by Prof. Illiapoulos, a student, and his nephew, who is just Carl's age.

I was surprised to find the weather here is just like in Pasadena, but about 5 degrees cooler: the vegetation is very similar, the hills look bare and desert-like—same plants, same cactuses, same low humidity and same cool nights. But there the similarity ends. Athens is a sprawling, ugly, noisy, exhaust-filled mess of streets filled with nervous traffic jumping like rabbits when the lights go green and stopping with squealing brakes when they go red—and blowing horns when they go yellow. Very similar to Mexico City, except the people don't look as poor—there are only

*Daughter Michelle was about eleven when this letter was written, in 1980 or 1981.

occasional beggars in the streets. You, Gweneth, would love it because there are so many shops (all small), and Carl would love walking around in the arcades with their rabbit-warren twists and surprises, especially in the old part of town.

Yesterday morning I went to the archeological museum. Michelle would like all the great Greek statues of horses—especially one of a small boy on a large galloping horse, all in bronze, that is a sensation. I saw so much stuff my feet began to hurt. I got all mixed up—things are not labeled well. Also, it was slightly boring because we have seen so much of that stuff before. Except for one thing: among all those art objects there was one thing so entirely different and strange that it is nearly impossible. It was recovered from the sea in 1900 and is some kind of machine with gear trains, very much like the inside of a modern wind-up alarm clock. The teeth are very regular and many wheels are fitted closely together. There are graduated circles and Greek inscriptions. I wonder if it is some kind of fake. There was an article on it in the Scientific American in 1959.

Yesterday afternoon I went to the Acropolis, which is right in the middle of the city—a high rock plateau on which was built the Parthenon and other shrines and temples. The Parthenon looks pretty good, but the Temple at Segesta, which Gweneth and I saw in Sicily, is just as impressive because you are allowed to walk around in it—you can't go up to or walk around among the Parthenon columns. Prof. Illiapoulos' sister came with us and with a notebook she had—she is a professional archeologist—guided our tour with all kinds of details, dates, quotations from Plutarch, etc.

It appears the Greeks take their past very seriously. They study ancient Greek archeology in their elementary schools for 6 years, having to take 10 hours of that subject

every week. It is a kind of ancestor worship, for they empha-
size always how wonderful the ancient Greeks were—and
wonderful indeed they were. When you encourage them by
saying, "Yes, and look how modern man has advanced
beyond the ancient Greeks"—thinking of experimental science, the development of mathematics, the art of the Renaissance, the great depth and understanding of the relative
shallowness of Greek philosophy, etc., etc.—they reply,
"What do you mean? What was wrong with the ancient
Greeks?" They continually put their age down and the old
age up, until to point out the wonders of the present seems
to them to be an unjustified lack of appreciation for the
past.

They were very upset when I said that the development of greatest importance to mathematics in Europe was
the discovery by Tartaglia that you can solve a cubic equa-
tion: although it is of very little use in itself, the discovery
must have been psychologically wonderful because it
showed that a modern man could do something no ancient
Greek could do. It therefore helped in the Renaissance,
which was the freeing of man from the intimidation of the
ancients. What the Greeks are learning in school is to be
intimidated into thinking they have fallen so far below their
super ancestors.

I asked the archeologist lady about the machine in the
museum—whether other similar machines, or simpler machines leading up to it or down from it, were ever found—
but she hadn't heard of it. So I met her and her son of Carl's
age (who looks at me as if I were a heroic ancient Greek,
for he is studying physics) at the museum to show it to her.
She required some explanation from me why I thought
such a machine was interesting and surprising because,
"Didn't Eratosthenes measure the distance to the sun, and
didn't that require elaborate scientific instruments?" Oh,
how ignorant are classically educated people. No wonder

they don't appreciate their own time. They are not of it and do not understand it. But after a bit she believed maybe it *was* striking, and she took me to the back rooms of the museum—surely there were other examples, and she would get a complete bibliography. Well, there were no other examples, and the complete bibiography was a list of three articles (including the one in the <u>Scientific American</u>) —all by one man, an <u>American</u> from Yale!

I guess the Greeks think all Americans must be dull, being only interested in machinery when there are all those beautiful statues and portrayals of lovely myths and stories of gods and goddesses to look at. (In fact, a lady from the museum staff remarked, when told that the professor from America wanted to know more about item 15087, "Of all the beautiful things in this museum, why does he pick out <u>that</u> particular item? What is so special about it?")

Everyone here complains of the heat, and concerned about whether you can stand it, when in fact it is just like Pasadena but about 5 degrees cooler on the average. So all stores and offices close from perhaps 1:30 pm to 5:30 pm ("because of the heat"). It turns out to be really a good idea (everyone takes a nap) because then they go late into the night—supper is between 9:30 and 10 pm, when it is cool. Right now, people here are seriously complaining about a new law: to save energy, all restaurants and taverns must close at 2 am. This, they say, will spoil life in Athens.

It is the witching hour between 1:30 and 5:30 pm now, and I am using it to write to you. I miss you, and I would really be happier at home. I guess I really have lost my bug for travelling. I have a day and a half yet here and they have given me all kinds of literature about a beautiful beach (of pebbles) here, of an important ancient site (although in rather complete ruins) there, etc. But I will go to none of them, for each, it turns out, is a long, two- to four-hour ride each way on a tour bus. No. I'll just stay here and prepare my talks for Crete. (They have me giving an extra three lectures to some twenty Greek university students who are all coming to Crete just to hear me. I'll do something like my New Zealand lectures,* but I haven't got any notes! I'll have to work them out again.)

I miss you all, especially when I go to bed at night—no dogs to scratch and say good night to!

Love, Richard.

P.S. IF YOU CAN'T READ THE ABOVE HANDWRITING, HAVE NO FEAR—IT IS UNIMPORTANT RAMBLINGS. I AM WELL & IN ATHENS.

MacFaddin Hall
Cornell University
Ithaca, NY
November 19, 1947†

My Dear Family:

Just a brief letter before we go off to Rochester. We have every Wednesday a seminar at which somebody talks about some item of research, and from time to time this is made a joint seminar with Rochester University. To-day is the first time this term that we are going over there for it.

It is a magnificent day, and it should be a lovely trip;

*The "New Zealand lectures," delivered in 1979, are written up in *QED: The Strange Theory of Light and Matter* (Princeton University Press, 1985).

†These letters were contributed by Freeman Dyson. They are the first and last letters he wrote that mention Richard Feynman. Other letters are referred to in Dyson's book *Disturbing the Universe*.

Rochester is northwest of here, on the shores of Lake Ontario, and we go through some wild country. I am being taken in Feynman's car, which will be great fun if we survive. Feynman is a man for whom I am developing a considerable admiration; he is the first example I have met of that rare species, the native American scientist. He has developed a private version of the quantum theory, which is generally agreed to be a good piece of work and may be more helpful than the orthodox version for some problems; in general he is always sizzling with new ideas, most of which are more spectacular than helpful, and hardly any of which get very far before some newer inspiration eclipses it. His most valuable contribution to physics is as a sustainer of morale; when he bursts into the room with his latest brain-wave and proceeds to expound on it with the most lavish sound effects and waving about of the arms, life at least is not dull.

Weisskopf, the chief theoretician at Rochester, is also an interesting and able man, but of the normal European type; he comes from Munich, where he was a friend of Bethe from student days.

The event of the last week has been a visit from Peierls, who . . . stayed two nights with the Bethes before flying home. . . . On Monday night the Bethes gave a party in his honor, to which most of the young theoreticians were invited. When we arrived we were introduced to Henry Bethe, who is now five years old, but he was not at all impressed. In fact, the only thing he would say was "I want Dick! You told me Dick was coming!" Finally he had to be sent off to bed, since Dick (alias Feynman) did not materialize.

About half an hour later, Feynman burst into the room, just had time to say, "So sorry I'm late—had a brilliant idea just as I was coming over," and then dashed upstairs to console Henry. Conversation then ceased while the company listened to the joyful sounds above, sometimes taking the form of a duet and sometimes of a one-man percussion band. . . .

Much Love,

Freeman

Urbana, Illinois
April 9, 1981

Dear Sara,*

I just spent a marvelous three days with Dick Feynman and wished you had been there to share him with us. Sixty years and a big cancer operation have not blunted him. He is still the same Feynman that we knew in the old days at Cornell.

We were together at a small meeting of physicists organized by John Wheeler at the University of Texas. For some reason Wheeler decided to hold the meeting at a grotesque place called World of Tennis, a country club where Texas oil-millionaires go to relax. So there we were. We all grumbled at the high prices and the extravagant ugliness of our rooms. But there was nowhere else to go—or so we thought. But Dick thought otherwise: he just said, "To hell with it. I am not going to sleep in this place," picked up his suitcase, and walked off alone into the woods.

In the morning he reappeared, looking none the worse for his night under the stars. He said he did not sleep much, but it was worth it.

We had many conversations about science and history, just like in the old days. But now he had something new to talk about, his children. He said, "I always thought I would be a specially good father because I wouldn't try to push my

*A family friend.

kids into any particular direction. I wouldn't try to turn
them into scientists or intellectuals if they didn't want it. I
would be just as happy with them if they decided to be truck
drivers or guitar players. In fact, I would even like it better
if they went out in the world and did something real instead
of being professors like me. But they always find a way to
hit back at you. My boy Carl, for instance. There he is in
his second year at MIT, and all he wants to do with his life
is to become a goddamn philosopher!"*

   As we sat in the airport waiting for our planes, Dick
pulled out a pad of paper and a pencil and started to draw
the faces of people sitting in the lounge. He drew them
amazingly well. I said I was sorry I have no talent for drawing. He said, "I always thought I have no talent either. But
you don't need any talent to do stuff like this." . . .

                                          Yours,

                                          Freeman

                                          February 17, 1988
                                          London, England†

Dear Mrs. Feynman,

   We have not met, I believe, frequently enough for
either of us to have taken root in the other's conscious
memory. So please forgive any impertinence, but I could
not let Richard's death pass unnoticed, or to take the opportunity to add my own sense of loss to yours.
   Dick was the best and favorite of several "uncles" who
encircled my childhood. During his time at Cornell he was

*As it turned out, Feynman was not to be disappointed: Carl works at the Think-
ing Machines Company, and daughter Michelle is studying to become a commer-
cial photographer.
†This letter was contributed by Henry Bethe.

a frequent and always welcome visitor at our house, one who could be counted on to take time out from conversations with my parents and other adults to lavish attention on the children. He was at once a great player of games with us and a teacher even then who opened our eyes to the world around us.

My favorite memory of all is of sitting as an eight- or nine-year-old between Dick and my mother, waiting for the distinguished naturalist Konrad Lorenz to give a lecture. I was itchy and impatient, as all young are when asked to sit still, when Dick turned to me and said, "Did you know that there are twice as many numbers as numbers?"

"No, there are not!" I was defensive as all young of my knowledge.

"Yes there are; I'll show you. Name a number."

"One million." A big number to start.

"Two million."

"Twenty-seven."

"Fifty-four."

I named about ten more numbers and each time Dick named the number twice as big. Light dawned.

"I see; so there are three times as many numbers as numbers."

"Prove it," said Uncle Dick. He named a number. I named one three times as big. He tried another. I did it again. Again.

He named a number too complicated for me to multiply in my head. "Three times that," I said.

"So, is there a biggest number?" he asked.

"No," I replied. "Because for every number, there is one twice as big, one three times as big. There is even one a million times as big."

"Right, and that concept of increase without limit, of no biggest number, is called 'infinity'."

At that point Lorenz arrived, so we stopped to listen to him.

I did not see Dick often after he left Cornell. But he left me with bright memories, infinity, and new ways of learning about the world. I loved him dearly.

Sincerely Yours,

Henry Bethe

Richard and Arlene on the boardwalk in Atlantic City.

On their wedding day.

Arlene in the hospital.

Coffee hour at Winnett Student Center, 1964. (CALTECH)

Gesturing at a Caltech Alumni Day lecture, 1978. (CALTECH)

In the Caltech production of Fiorello, *1978. (CALTECH)*

The chief from Bali Hai in South Pacific, *1982. (CALTECH)*

Describing Feynman diagrams, 1984. (FAUSTIN BRAY)

Modulating sounds of the "crazy drum" with Ralph Leighton, 1984. (FAUSTIN BRAY)

With Michelle, 3, and Carl, 10, in Yorkshire, England. (BBC,
YORKSHIRE TELEVISION)

*With son Carl on the day
Richard won the Nobel
Prize, 1965.*
(CALTECH)

Richard and Gweneth on their silver wedding anniversary, 1985.
(PHOTO BY YASUSHI OHNUKI)

Richard Feynman began taking art lessons at the age of 44, and continued drawing for the rest of his life. These sketches include portraits of professional models, his friend Bob Sadler, and his daughter Michelle (at the age of 14). Feynman signed all of his artwork "Ofey" to make sure no one would suspect who really drew them.

Dot Sadler

Part 2

MR. FEYNMAN GOES
TO WASHINGTON:
INVESTIGATING
THE SPACE SHUTTLE
CHALLENGER
DISASTER

Preliminaries

IN THIS STORY I'm going to talk a lot about NASA,* but when I say "NASA did this" and "NASA did that," I don't mean all of NASA; I just mean that part of NASA associated with the shuttle.

To remind you about the shuttle, the large central part is the tank, which holds the fuel: liquid oxygen is at the top, and liquid hydrogen is in the main part. The engines which burn that fuel are at the back end of the orbiter, which goes into space. The crew sits in the front of the orbiter; behind them is the cargo bay.

During the launch, two solid-fuel rockets boost the shuttle for a few minutes before they separate and fall back into the sea. The tank separates from the orbiter a few minutes later—much higher in the atmosphere—and breaks up as it falls back to earth.

The solid rocket boosters are made in sections. There are two types of joints to hold the sections together: the permanent "factory joints" are sealed at the Morton Thiokol factory in Utah; the temporary "field joints" are sealed before each flight—"in the field"—at the Kennedy Space Center in Florida.

*The National Aeronautics and Space Administration.

FIGURE 1. *The space shuttle* Challenger. *The fuel tank, flanked by two solid-fuel rocket boosters, is attached to the orbiter, whose main engines burn liquid hydrogen and liquid oxygen. (© NASA.)*

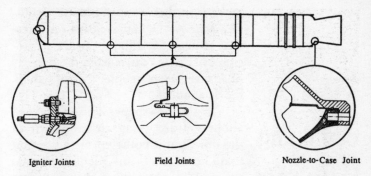

Igniter Joints   Field Joints   Nozzle-to-Case Joint

FIGURE 2. Locations and close-up views of booster-rocket field joints.

Committing
Suicide

AS YOU probably know, the space shuttle *Challenger* had an accident on Tuesday, January 28, 1986. I saw the explosion on TV, but apart from the tragedy of losing seven people, I didn't think much about it.

In the newspaper I used to read about shuttles going up and down all the time, but it bothered me a little bit that I never saw in any scientific journal any results of anything that had ever come out of the experiments on the shuttle that were supposed to be so important. So I wasn't paying very much attention to it.

Well, a few days after the accident, I get a telephone call from the head of NASA, William Graham, asking me to be on the committee investigating what went wrong with the shuttle! Dr. Graham said he had been a student of mine at Caltech, and later had worked at the Hughes Aircraft Company, where I gave lectures every Wednesday afternoon.

I still wasn't exactly sure who he was.

When I heard the investigation would be in Washington, my immediate reaction was not to do it: I have a principle of not going anywhere near Washington or having anything to do with government, so my immediate reaction was—how am I gonna get out of this?

I called various friends like Al Hibbs and Dick Davies, but they explained to me that investigating the *Challenger* accident was very important for the nation, and that I should do it.

My last chance was to convince my wife. "Look," I said. "Anybody could do it. They can get somebody else."

"No," said Gweneth. "If you don't do it, there will be twelve people, all in a group, going around from place to place together. But if you join the commission, there will be eleven people—all in a group, going around from place to place together—while the twelfth one runs around all over the place, checking all kinds of unusual things. There probably won't be anything, but if there is, you'll find it." She said, "There isn't anyone else who can do that like you can."

Being very immodest, I believed her.

Well, it's one thing to figure out what went wrong with the shuttle. But the next thing would be to find out what was the matter with the organization of NASA. Then there are questions like, "Should we continue with the shuttle system, or is it better to use expendable rockets?" And then come even bigger questions: "Where do we go from here?" "What should be our future goals in space?" I could see that a commission which started out trying to find out what happened to the shuttle could end up as a commission trying to decide on national policy, and go on forever!

That made me quite nervous. I decided to get out at the end of six months, no matter what.

But I also resolved that while I was investigating the accident, I shouldn't do anything else. There were some physics problems I was playing with. There was a computer class at Caltech I was teaching with another professor. (He offered to take over the course.) There was the Thinking Machines Company in Boston I was going to consult for.

(They said they would wait.) My physics would have to wait, too.

By this time it was Sunday. I said to Gweneth, "I'm gonna commit suicide for six months," and picked up the telephone.

WHEN I called Graham and accepted, he didn't know exactly what the commission was going to do, who it was going to be under, or even if I would be accepted onto it. (There was still hope!)

But the next day, Monday, I got a telephone call at 4 P.M.: "Mr. Feynman, you have been accepted onto the commission"—which by that time was a "presidential commission" headed by William P. Rogers.

I remembered Mr. Rogers. I felt sorry for him when he was secretary of state, because it seemed to me that President Nixon was using the national security adviser (Kissinger) more and more, to the point where the secretary of state was not really functioning.

At any rate, the first meeting would be on Wednesday. I figured there's nothing to do on Tuesday—I could fly to Washington Tuesday night—so I called up Al Hibbs and asked him to get some people at JPL* who know something about the shuttle project to brief me.

On Tuesday morning I rush over to JPL, full of steam, ready to roll. Al sits me down, and different engineers come in, one after the other, and explain the various parts of the shuttle. I don't know *how* they knew, but they knew all about the shuttle. I got a very

The Cold Facts

*NASA's Jet Propulsion Laboratory, located in Pasadena; it is administered by Caltech.

R. P. FEYNMAN

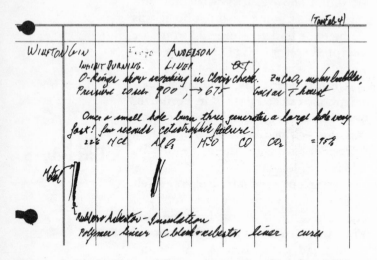

FIGURE 3. *The beginning of Feynman's notes from his informal JPL briefing.*

thorough, high-speed, intense briefing. The guys at JPL had the same enthusiasm that I did. It was really quite exciting.

When I look at my notes now, I see how quickly they gave me hints about where to look for the shuttle's problems. The first line of my notes says "Inhibit burning. Liner." (To inhibit propellant from burning through the metal wall of each booster rocket, there's a liner, which was not working right.) The second line of my notes says "O-rings show scorching in clevis check." It was noticed that hot gas occasionally burned past the O-rings in booster-rocket field joints.

On the same line it says "Zn CrO$_4$ makes bubbles." (The zinc chromate putty, packed as an insulator behind the O-rings, makes bubbles which can become enlarged

The Cold Facts   121

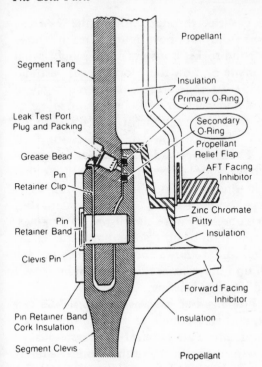

FIGURE 4. Detailed diagram of a field joint.

FIGURE 5. Photograph of bubbles in zinc chromate putty, which can lead to erosion of the O-rings.

very fast when hot gas leaks through, eroding the O-rings.)

The engineers told me how much the pressures change inside the solid rocket boosters during flight, what the propellant is made of, how the propellant is cast and then baked at different temperatures, the percentages of asbestos, polymers, and whatnot in the liner, and all kinds of other stuff. I learned about the thrusts and forces in the engines, which are the most powerful engines for their weight ever built. The engines had many difficulties, especially cracked turbine blades. The engineers told me that some of the people who worked on the engines always had their fingers crossed on each flight, and the moment they saw the shuttle explode, they were sure it was the engines.

If the engineers didn't know something, they'd say something like, "Oh, Lifer knows about that; let's get *him* in." Al would call up Lifer, who would come right away. I couldn't have had a better briefing.

It's called a briefing, but it wasn't brief: it was *very* intense, very fast, and very complete. It's the only way I know to get technical information quickly: you don't just sit there while they go through what *they* think would be interesting; instead, you ask a lot of questions, you get quick answers, and soon you begin to understand the circumstances and learn just what to ask to get the next piece of information you need. I got one hell of a good education that day, and I sucked up the information like a sponge.

That night I took the red-eye* to Washington, and got there early Wednesday morning. (I never took the red-eye again—I learned!)

I checked into the Holiday Inn in downtown Washington, and got a cab to take me to the first meeting of the commission.

"Where to?" the driver says.

---

*Note for foreign readers: a flight that leaves the West Coast around 11 P.M. and arrives on the East Coast around 7 A.M., five hours and three time zones later.

All I have is a little piece of paper. "1415 8th Street."

We start off. I'm new in Washington. The Capitol is over here, the Washington Monument is over there; everything seems very close. But the taxi goes on and on, farther and farther into worse and worse territory. Buildings get smaller, and they begin to look run down a little bit. Finally, we get onto 8th Street, and as we go along, the buildings begin to disappear altogether. Finally we find the address—by interpolation: it's an empty lot between two buildings!

By this time I realize something is completely cockeyed. I don't know what to do, because I've only got this slip of paper, and I don't know where to go.

I say to the taxi driver, "The meeting I'm going to has something to do with NASA. Can you take me to NASA?"

"Sure," he says. "You know where it is, don't you? It's right where I picked you up!"

It was true. NASA I could have walked to from the Holiday Inn: it was right across the street! I go in, past the guard at the gate, and start wandering around.

I find my way to Graham's office, and ask if there's a meeting about the shuttle.

"Yes, I know where it is," somebody says. "I'll take you down there."

They take me to a room and, sure enough, there's a big meeting going on: there are bright lights and television cameras down in front; the room is completely full, bursting with people, and all I can do is barely squash my way into the back. I'm thinking, "There's only one door to this place. How the hell am I gonna get down to the front from here?"

Then I overhear something a little bit—it's so far down there that I can't make out exactly what it is—but it's evidently a different subject!

So I go back to Graham's office and find his secretary. She calls around and finds out where the commission is meeting. "I don't know, either," she says to the person on the other end. "He simply wandered in here!"

The meeting was in Mr. Rogers' law offices, at 1415 *H* Street. My slip of paper said 1415 *8th* Street. (The address had been given over the telephone.)

I finally got to Mr. Rogers' office—I was the only one late—and Mr. Rogers introduced me to the other commissioners. The only one I had ever heard of besides Mr. Rogers was Neil Armstrong, the moon man, who was serving as vice-chairman. (Sally Ride was on the commission, but I didn't realize who she was until later.*) There was a very handsome-looking guy in a uniform, a General Kutyna (pronounced Koo-TEE-na). He looked formidable in his outfit, while the other people had on ordinary suits.

This first meeting was really just an informal get-together. That bothered me, because I was still wound up like a spring from my JPL briefing the day before.

Mr. Rogers did announce a few things. He read from the executive order that defined our work:

The Commission shall:
1. Review the circumstances surrounding the accident and establish the probable cause or causes of the accident; and
2. Develop recommendations for corrective or other action based upon the Commission's findings and determinations.

Mr. Rogers also said we would complete our investigation within 120 days.

That was a relief: the scope of our commission would be limited to investigating the accident, and our work might be finished before I was done committing suicide!

Mr. Rogers asked each of us how much of our time we could spend on the commission. Some of the commissioners were retired, and almost everybody said they had rearranged their schedules. I said, "I'm ready to work

*Note for foreign readers: Sally Ride was the first American woman in space.

100 percent, starting right now!"

Mr. Rogers asked, "Who will be in charge of writing the report?"

A Mr. Hotz, who had been the editor of *Aviation Week* magazine, volunteered to do that.

Then Mr. Rogers brought up another matter. "I've been in Washington a long time," he said, "and there's one thing you all must know: no matter what we do, there will always be leaks to the press. The best we can do is just try to minimize them. The proper way to deal with leaks is to have public meetings. We will have closed meetings, of course, but if we find anything important, we will have an open meeting right away, so the public will always know what is going on."

Mr. Rogers continued, "To start things off right with the press, our first official meeting will be a public meeting. We'll meet tomorrow at 10 A.M."

As we were leaving the get-together, I heard General Kutyna say, "Where's the nearest Metro station?"

I thought, "This guy, I'm gonna get along with him fine: he's dressed so fancy, but inside, he's straight. He's not the kind of general who's looking for his driver and his special car; he goes back to the Pentagon by the Metro." Right away I liked him, and over the course of the commission I found my judgment in this case was excellent.

The next morning, a limousine called for me—someone had arranged for us to arrive at our first official meeting in limousines. I sat in the front seat, next to the driver.

On the way to the meeting, the driver says to me, "I understand a lot of important people are on this commission . . ."

"Yeah, I s'pose . . ."

"Well, I collect autographs," he says. "Could you do me a favor?"

"Sure," I say.

I'm reaching for my pen when he says, "When we get there, could you point out to me which one Neil Armstrong is, so I can get his autograph?"

Before the meeting started, we were sworn in. People were milling around; a secretary handed us each a badge with our picture on it so we could go anywhere in NASA. There were also some forms to sign, saying you agree to this and that so you can get your expenses paid, and so on.

After we were sworn in, I met Bill Graham. I did recognize him, and remembered him as a nice guy.

This first public meeting was going to be a general briefing and presentation by the big cheeses of NASA—Mr. Moore, Mr. Aldrich, Mr. Lovingood, and others. We were seated in big leather chairs on a dais, and there were bright lights and TV cameras pointing at us every time we scratched our noses.

I happened to sit next to General Kutyna. Just before the meeting started, he leans over and says, "Co-pilot to pilot: comb your hair."

I say, "Pilot to co-pilot: can I borrow your comb?"

The first thing we had to learn was the crazy acronyms that NASA uses all over the place: "SRMs" are the solid rocket motors, which make up most of the "SRBs," the solid rocket boosters. The "SSMEs" are the space shuttle main engines; they burn "LH" (liquid hydrogen) and "LOX" (liquid oxygen), which are stored in the "ET," the external tank. Everything's got letters.

And not just the big things: practically every valve has an acronym, so they said, "We'll give you a dictionary for the acronyms—it's really very simple." Simple, sure, but the dictionary is a great, big, fat book that you've gotta keep looking through for things like "HPFTP" (high-pressure fuel turbopump) and "HPOTP" (high-pressure oxygen turbopump).

Then we learned about "bullets"—little black circles in front of phrases that were supposed to summarize

STS 51-L CARGO ELEMENTS

● TRACKING AND DATA RELAY SATELLITE-B/INERTIAL UPPER STAGE

● SPARTAN-HALLEY/MISSION PECULIAR SUPPORT STRUCTURE

● CREW COMPARTMENT
  - TISP - TEACHER IN SPACE PROGRAM
  - CHAMP - COMET HALLEY ACTIVE MONITORING PROGRAM
  - FDE - FLUID DYNAMICS EXPERIMENT
  - STUDENT EXPERIMENTS
  - RME - RADIATION MONITORING EXPERIMENT
  - PPE - PHASE PARTITIONING EXPERIMENT

FIGURE 6. An example of "bullets."

things. There was one after another of these little goddamn bullets in our briefing books and on the slides.

It turned out that apart from Mr. Rogers and Mr. Acheson, who were lawyers, and Mr. Hotz, who was an editor, we all had degrees in science: General Kutyna had a degree from MIT; Mr. Armstrong, Mr. Covert, Mr. Rummel, and Mr. Sutter were all aeronautical engineers, while Ms. Ride, Mr. Walker, Mr. Wheelon, and I were all physicists. Most of us seemed to have done some preliminary work on our own. We kept asking questions that were much more technical than some of the big cheeses were prepared for.

When one of them couldn't answer a question, Mr. Rogers would reassure him that we understood he wasn't expecting such detailed questions, and that we were satisfied, for the time being at least, by the perpetual answer, "We'll get that information to you later."

The main thing I learned at that meeting was how inefficient a public inquiry is: most of the time, other people are asking questions you already know the answer to—or are not interested in—and you get so fogged out that you're hardly listening when important points are being passed over.

What a contrast to JPL, where I had been filled with all sorts of information very fast. On Wednesday we have a "get-together" in Mr. Rogers' office—that takes two hours—and then we've got the rest of the day to do what? Nothing. And that night? Nothing. The next day, we have the public meeting—"We'll get back to you on that"—which equals nothing! Although it *looked* like we were doing something every day in Washington, we were, in reality, sitting around doing nothing most of the time.

That night I gave myself something to do: I wrote out the kinds of questions I thought we should ask during our investigation, and what topics we should study. My plan was to find out what the rest of the commission wanted to do, so we could divide up the work and get going.

The next day, Friday, we had our first real meeting. By this time we had an office—we met in the Old Executive Office Building—and there was even a guy there to transcribe every word we said.

Mr. Rogers was delayed for some reason, so while we waited for him, General Kutyna offered to tell us what an accident investigation is like. We thought that was a good idea, so he got up and explained to us how the air force had proceeded with its investigation of an unmanned Titan rocket which had failed.

I was pleased to see that the system he described—what the questions were, and the way they went about finding the answers—was very much like what I had laid out the night before, except that it was much more methodical than I had envisioned. General Kutyna warned us that sometimes it looks like the cause is obvious, but when you investigate more carefully you have to change your mind. They had very few clues, and changed their minds three times in the case of the Titan.

I'm all excited. I want to do this kind of investigation, and figure we can get started right away—all we have to do is decide who will do what.

But Mr. Rogers, who came in partway through General Kutyna's presentation, says, "Yes, your investigation was a great success, General, but we won't be able to use your methods here because we can't get as much information as you had."

Perhaps Mr. Rogers, who is not a technical man, did not realize how patently false that was. The Titan, being an unmanned rocket, didn't have anywhere near the number of check gadgets the shuttle did. We had television pictures showing a flame coming out the side of a booster rocket a few seconds before the explosion; all we could see in General Kutyna's pictures of the Titan was a lousy dot in the sky—just a little, tiny flash—and he was able to figure stuff out from that.

Mr. Rogers says, "I have arranged for us to go to Florida next Thursday. We'll get a briefing there from NASA officials, and they'll take us on a tour of the Kennedy Space Center."

I get this picture of the czarina coming to a Potemkin village: everything is all arranged; they show us how the rocket looks and how they put it together. It's not the way to find out how things *really* are.

Then Mr. Armstrong says, "We can't expect to do a technical investigation like General Kutyna did." This bothered me a lot, because the only things I pictured myself doing were technical! I didn't know exactly what he meant: perhaps he was saying that all the technical lab work would be done by NASA.

I began suggesting things I could do.

While I'm in the middle of my list, a secretary comes in with a letter for Mr. Rogers to sign. In the interim, when I've just been shut up and I'm waiting to come back, various other commission members offer to work with me. Then Mr. Rogers looks up again to continue the meeting, but he calls on somebody else—as if he's absentminded and forgot I'd been interrupted. So I have to get the floor again, but when I start my stuff again, another "accident" happens.

In fact, Mr. Rogers brought the meeting to a close while I was in midstream! He repeated his worry that we'll never really figure out what happened to the shuttle.

This was extremely discouraging. It's hard to understand now, because NASA has been taking at least two years to put the shuttle back on track. But at the time, I thought it would be a matter of days.

I went over to Mr. Rogers and said, "We're going to Florida next Thursday. That means we've got nothing to do for *five days:* what'll I do for five days?"

"Well, what would you have done if you hadn't been on the commission?"

"I was going to go to Boston to consult, but I canceled it in order to work 100 percent."

"Well, why don't you go to Boston for the five days?"

I couldn't take that. I thought, "I'm dead already! The goddamn thing isn't working right." I went back to my hotel, devastated.

Then I thought of Bill Graham, and called him up. "Listen, Bill," I said. "You got me into this; now you've gotta save me: I'm completely depressed; I can't stand it."

He says, "What's the matter?"

"I want to *do* something! I want to go around and talk to some engineers!"

He says, "Sure! Why not? I'll arrange a trip for you. You can go wherever you want: you could go to Johnson, you could go to Marshall, or you could go to Kennedy . . ."

I thought I wouldn't go to Kennedy, because it would look like I'm rushing to find out everything ahead of the others. Sally Ride worked at Johnson, and had offered to work with me, so I said, "I'll go to Johnson."

"Fine," he says. "I'll tell David Acheson. He's a personal friend of Rogers, and he's a friend of mine. I'm sure everything will be okay."

Half an hour later, Acheson calls me: "I think it's a great idea," he says, "and I told Mr. Rogers so, but he says no. I just don't know why I can't convince him."

Meanwhile, Graham thought of a compromise: I would stay in Washington, and he would get people to come to his office at NASA, right across the street from my hotel. I would get the kind of briefing I wanted, but I wouldn't be running around.

Then Mr. Rogers calls me: he's against Graham's compromise. "We're all going to Florida next Thursday," he says.

I say, "If the idea is that we sit and listen to briefings, it won't work with me. I can work much more efficiently if I talk to engineers directly."

"We have to proceed in an orderly manner."

"We've had several meetings by now, but we still haven't been assigned anything to do!"

Rogers says, "Well, do you want me to bother all the other commissioners and call a special meeting for Monday, so we can make such assignments?"

"Well, yes!" I figured our job was to work, and we *should* be bothered—you know what I mean?

So he changes the subject, naturally. He says, "I understand you don't like the hotel you're in. Let me put you in a good hotel."

"No, thank you; everything is fine with my hotel."

Pretty soon he tries again, so I say, "Mr. Rogers, my personal comfort is not what I'm concerned with. I'm trying to get to work. I want to *do* something!"

Finally, Rogers says it's okay to go across the street to talk to people at NASA.

I was obviously quite a pain in the ass for Mr. Rogers.

Later, Graham tried to explain it to me. "Suppose you, as a technical person, were given the job as chairman of a committee to look into some legal question. Your commission is mostly lawyers, and one of them keeps saying, 'I can work more effectively if I talk directly to other lawyers.' I assume you'd want to get your bearings first, before letting anybody rush off investigating on his own."

Much later, I appreciated that there were lots of problems which Mr. Rogers had to address. For example, any piece of information any of us received had to be entered into the record and made available to the other commissioners, so a central library had to be set up. Things like that took time.

On Saturday morning I went to NASA. Graham brought in guys to tell me all about the shuttle. Although they were pretty high up in NASA, the guys were technical.

The first guy told me all about the solid rocket boosters—the propellant, the motor, the whole thing except the seals. He said, "The seals expert will be here this afternoon."

The next guy told me all about the engine. The basic operation was more or less straightforward, but then there were all kinds of controls, with backing and hauling from pipes, heating from this and that, with high-pressure hydrogen pushing a little propeller which turns something else, which pumps oxygen through a vent valve—that kind of stuff.

It was interesting, and I did my best to understand it, but after a while I told the fella, "That's as much as I'm going to take, now, on the engine."

"But there are many problems with the engines that you should hear about," he says.

I was hot on the trail of the booster rocket, so I said, "I'll have to put off the main engines till later, when I have more time."

Then a guy came in to tell me about the orbiter. I felt terrible, because he had come in on a Saturday to see me, and it didn't look like the orbiter had anything to do with the accident. I was having enough trouble understanding the rest of the shuttle—there's only a certain amount of information per cubic inch a brain can hold—so I let him tell me some of the stuff, but soon I had to tell him that it was getting too detailed, so we just had a pleasant conversation.

In the afternoon, the seals expert came in—his name was Mr. Weeks—and gave me what amounted to a continuation of my JPL briefing, with still more details.

There's putty and other things, but the ultimate seal is supposed to be two rubber rings, called O-rings, which are approximately a quarter of an inch thick and lie on a circle 12 feet in diameter—that's something like 37 feet long.

When the seals were originally designed by the Morton Thiokol Company, it was expected that pressure from the burning propellant would squash the O-rings. But because the joint is stronger than the wall (it's three times thicker), the wall bows outward, causing the joint to bend a little—enough to lift the rubber O-rings off the seal area. Mr. Weeks told me this phenomenon is called "joint rotation," and it was discovered very early, before they ever flew the shuttle.

The pieces of rubber in the joints are called O-rings, but they're not used like normal O-rings are. In ordinary circumstances, such as sealing oil in the motor of an automobile, there are sliding parts and rotating shafts, but the gaps are always the same. An O-ring just sits there, in a fixed position.

But in the case of the shuttle, the gap *expands* as the pressure builds up in the rocket. And to maintain the seal, the rubber has to expand *fast* enough to close the gap—and during a launch, the gap opens in a fraction of a second.

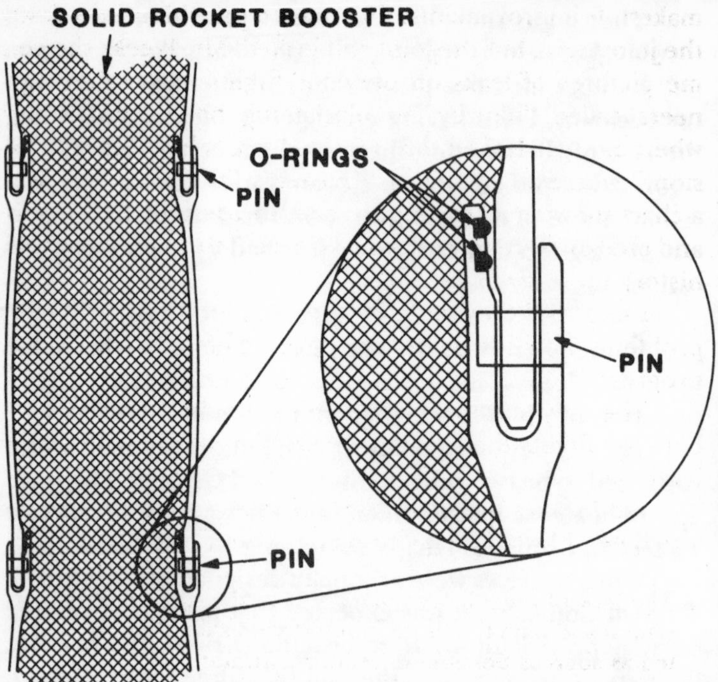

SOLID ROCKET BOOSTER

FIGURE 7. Joint rotation is caused by pressure from inside the rocket pushing the walls out farther than the joints. A gap opens, and hot gas flows past one or both of the O-rings.

Thus the resilience of the rubber became a very essential part of the design.

When the Thiokol engineers were discovering these problems, they went to the Parker Seal Company, which manufactures the rubber, to ask for advice. The Parker Seal Company told Thiokol that O-rings are not meant to be used that way, so they could give no advice.

Although it was known from nearly the beginning that the joint was not working as it was designed to, Thiokol kept struggling with the device. They made a number of makeshift improvements. One was to put shims in to keep the joint tight, but the joint still leaked. Mr. Weeks showed me pictures of leaks on previous flights—what the engineers called "blowby," a blackening behind an O-ring where hot gas leaked through, and what they called "erosion," where an O-ring had burned a little bit. There was a chart showing all the flights, and how serious the blowby and erosion were on each one. We went through the whole history up to *the* flight, 51-L.

I said, "Where does it say they were ever discussing the problem—how it's going along, or whether there's some progress?"

The only place was in the "flight readiness reviews"—between flights there was no discussion of the seals problem!

We looked at the summary of the report. Everything was behind little bullets, as usual. The top line says:

- The lack of a good secondary seal in the field joint is most critical and ways to reduce joint rotation should be incorporated as soon as possible to reduce criticality.

And then, near the bottom, it says:

- Analysis of existing data indicates that it is safe to continue flying existing design as long as all joints are leak checked* with a 200 psig stabilization . . .

I was struck by the contradiction: "If it's 'most critical,' how could it be 'safe to continue flying'? What's the *logic* of this?"

Mr. Weeks says, "Yes, I see what you mean! Well, let's see: it says here, 'Analysis of existing data . . .' "

We went back through the report and found the analysis. It was some kind of computer model with various as-

---

*Later in our investigation we discovered that it was this leak check which was a likely cause of the dangerous bubbles in the zinc chromate putty that I had heard about at JPL.

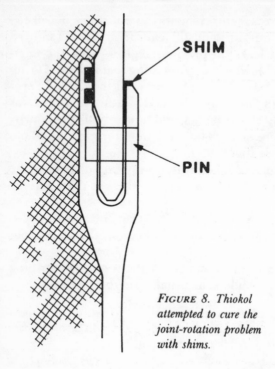

FIGURE 8. *Thiokol attempted to cure the joint-rotation problem with shims.*

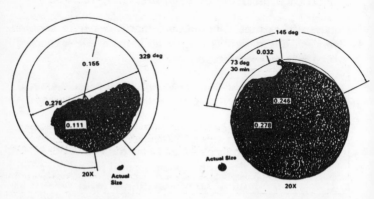

FIGURE 9. *Two examples of O-ring erosion. Such erosion would occur unpredictably along 2 or 3 inches of the 37-foot O-ring.*

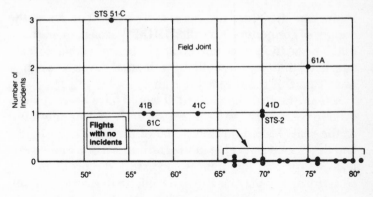

FIGURE 10. The correlation between temperature and O-ring incidents.

Recommendations

- The lack of a good secondary seal in the field joint is most critical and ways to reduce joint rotation should be incorporated as soon as possible to reduce criticality

- The flow conditions in the joint areas during ignition and motor operation need to be established through cold flow modeling to eliminate O-ring erosion

- QM-5 static test should be used to qualify a second source of the only flight certified joint filler material (asbestos-filled vacuum putty) to protect the flight program schedule

- VLS-1 should use the only flight certified joint filler material (Randolph asbestos-filled vacuum putty) in all joints

- Additional hot and cold subscale tests need to be conducted to improve analytical modeling of O-ring erosion problem and for establishing margins of safety for eroded O-rings

- Analysis of existing data indicates that it is safe to continue flying existing design as long as all joints are leak checked with a 200 psig stabilization pressure, are free of contamination in the seal areas and meet O-ring squeeze requirements

- Efforts needs to continue at an accelerated pace to eliminate SRM seal erosion

FIGURE 11. The self-contradictory recommendations of the seals report are underlined.

sumptions that were not necessarily right. You know the
danger of computers, it's called GIGO: garbage in, garbage
out! The analysis concluded that a little unpredictable leak-
age here and there could be tolerated, even though it
wasn't part of the original design.

If *all* the seals had leaked, it would have been obvious
even to NASA that the problem was serious. But only a few
of the seals leaked on only some of the flights. So NASA
had developed a peculiar kind of attitude: if one of the seals
leaks a little and the flight is successful, the problem isn't
so serious. Try playing Russian roulette that way: you pull
the trigger and the gun doesn't go off, so it must be safe
to pull the trigger again . . .

Mr. Weeks said there was a rumor that the history of
the seals problem was being leaked to the newspapers.
That bothered him a little bit, because it made NASA look
like it was trying to keep things secret.

I told him I was entirely satisfied with the people Graham had brought in to talk to me, and that since I had
already heard about the seals problem at JPL, it wasn't any
big deal.

The next day, Sunday, Bill Graham took me with his
family to the National Air and Space Museum. We had an
early breakfast together, and then we went across the street
to the museum.

I was expecting to see big crowds there, but I had
forgotten that Graham was such a big shot. We had the
whole place to ourselves for a while.

We did see Sally Ride there. She was in a display case,
in an astronaut's suit, holding a helmet and everything. The
wax model looked exactly like her.

At the museum there was a special theater with a movie
about NASA and its achievements. The movie was wonderful. I had not fully appreciated the enormous number of

people who were working on the shuttle, and all the effort that had gone into making it. And you know how a movie is: they can make it dramatic. It was so dramatic that I almost began to cry. I could see that the accident was a terrible blow. To think that so many people were working so hard to make it go—and then it busts—made me even more determined to help straighten out the problems of the shuttle as quickly as possible, to get all those people back on track. After seeing this movie I was very changed, from my semi anti-NASA attitude to a very strong pro-NASA attitude.

That afternoon, I got a telephone call from General Kutyna.

"Professor Feynman?" he says. "I have some urgent news for you. Uh, just a minute."

I hear some military-type band music in the background.

The music stops, and General Kutyna says, "Excuse me, Professor; I'm at an Air Force Band concert, and they just played the national anthem."

I could picture him in his uniform, standing at attention while the band is playing the "Star Spangled Banner," saluting with one hand and holding the telephone with the other. "What's the news, General?"

"Well, the first thing is, Rogers told me to tell you not to go over to NASA."

I didn't pay any attention to that, because I had already gone over to NASA the day before.

He continued, "The other thing is, we're going to have a special meeting tomorrow afternoon to hear from a guy whose story came out in the *New York Times* today."

I laughed inside: so we're going to have a special meeting on Monday, anyway!

Then he says, "I was working on my carburetor this morning, and I was thinking: the shuttle took off when the temperature was 28 or 29 degrees. The coldest temperature previous to that was 53 degrees. You're a professor; *what, sir, is the effect of cold on the O-rings?"*

"Oh!" I said. "It makes them stiff. Yes, of course!"

That's all he had to tell me. It was a clue for which I got a lot of credit later, but it was his observation. A professor of theoretical physics always has to be told what to look for. He just uses his knowledge to explain the observations of the experimenters!

On Monday morning General Kutyna and I went over to Graham's office and asked him if he had any information on the effects of temperature on the O-rings. He didn't have it on hand, but said he would get it to us as soon as possible.

Graham did, however, have some interesting photographs to show us. They showed a flame growing from the right-hand solid rocket booster a few seconds before the explosion. It was hard to tell exactly where the flame was coming out, but there was a model of the shuttle right there in the office. I put the model on the floor and walked around it until it looked exactly like the picture—in size, and in orientation.

I noticed that on each booster rocket there's a little hole—called the leak test port—where you can put pressure in to test the seals. It's *between* the two O-rings, so if it's not closed right and if the first O-ring fails, the gas would go out through the hole, and it would be a catastrophe. It was just about where the flame was. Of course, it was still a question whether the flame was coming out of the leak test port or a larger flame was coming out farther around, and we were seeing only the tip of it.

That afternoon we had our emergency closed meeting to hear from the guy whose story was in the *New York Times*. His name was Mr. Cook. He was in the budget department of NASA when he was asked to look into a possible seals problem and to estimate the costs needed to rectify it.

By talking to the engineers, he found out that the seals had been a big problem for a long time. So he reported that it would cost so-and-so much to fix it—a lot of money. From the point of view of the press and some of the commissioners, Mr. Cook's story sounded like a big exposé, as if NASA was hiding the seals problem from us.

I had to sit through this big, unnecessary excitement, wondering if every time there was an article in the newspaper, would we have to have a special meeting? We would never get anywhere that way!

But later, during that same meeting, some very interesting things happened. First, we saw some pictures which showed puffs of smoke coming out of a field joint just after ignition, before the shuttle even got off the pad. The smoke was coming out of the same place—possibly the leak test port—where the flame appeared later. There wasn't much question, now. It was all fitting together.

Then something happened that was completely unexpected. An engineer from the Thiokol Company, a Mr. MacDonald, wanted to tell us something. He had come to our meeting on his own, uninvited. Mr. MacDonald reported that the Thiokol engineers had come to the conclusion that low temperatures had something to do with the seals problem, and they were very, very worried about it. On the night before the launch, during the flight readiness review, they told NASA the shuttle shouldn't fly if the temperature was below 53 degrees—the previous lowest temperature—and on that morning it was 29.

Mr. MacDonald said NASA was "appalled" by that statement. The man in charge of the meeting, a Mr. Mulloy, argued that the evidence was "incomplete"—some flights with erosion and blowby had occurred at *higher* than 53 degrees—so Thiokol should reconsider its opposition to flying.

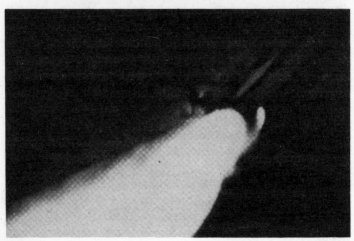

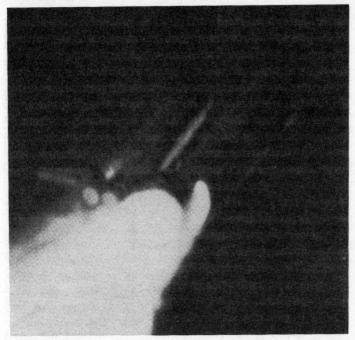

FIGURE 12. *Progression of a flame, possibly from the leak test port area. (© NASA.)*

Thiokol reversed itself, but MacDonald refused to go along, saying, "If something goes wrong with this flight, I wouldn't want to stand up in front of a board of inquiry and say that I went ahead and told them to go ahead and fly this thing outside what it was qualified to."

That was so astonishing that Mr. Rogers had to ask, "Did I understand you correctly, that you said . . . ," and he repeated the story. And MacDonald says, "Yes, sir."

The whole commission was shocked, because this was the first time any of us had heard *this* story: not only was there a failure in the seals, but there may have been

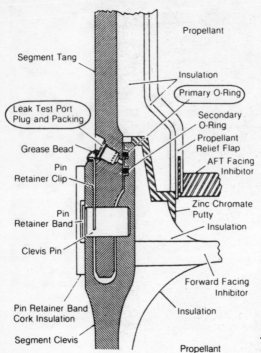

Figure 13. An incorrectly sealed leak test port could provide an escape route for a flame which burns past the primary O-ring.

a failure in management, too.

Mr. Rogers decided that we should look carefully into Mr. MacDonald's story, and get more details before we made it public. But to keep the public informed, we would have an open meeting the following day, Tuesday, in which Mr. Cook would testify.

I thought, "This is going to be like an act: we're going to say the same things tomorrow as we did today, and we won't learn anything new."

As we were leaving, Bill Graham came over with a stack of papers for me.

"Geez! That's fast!" I said. "I only asked you for the information this morning!" Graham was always very cooperative.

FIGURE 14. Puffs of black "smoke" (fine, unburned particles) were seen escaping from the same place where the flame was observed. (© NASA.)

The paper on top says, "Professor Feynman of the Presidential Commission wants to know about the effects over time of temperature on the resiliency of the O-rings . . ."—it's a memorandum addressed to a subordinate.

Under that memo is another memo: "Professor Feynman of the Presidential Commission wants to know . . ."—from that subordinate to *his* subordinate, and so on down the line.

There's a paper with some numbers on it from the poor bastard at the bottom, and then there's a series of submission papers which explain that the answer is being sent up to the next level.

So here's this stack of papers, just like a sandwich, and in the middle is the answer—to the wrong question! The answer was: "You squeeze the rubber for two hours at a certain temperature and pressure, and then see how long it takes to creep back"—over *hours*. I wanted to know how fast the rubber responds in *milliseconds* during a launch. So the information was of no use.

I went back to my hotel. I'm feeling lousy and I'm eating dinner; I look at the table, and there's a glass of ice water. I say to myself, "Damn it, *I* can find out about that rubber *without* having NASA send notes back and forth: I just have to *try* it! All I have to do is get a sample of the rubber."

I think, "I could do this tomorrow while we're all sittin' around, listening to this Cook crap we heard today. We always get ice water in those meetings; that's something I can do to save time."

Then I think, "No, that would be gauche."

But then I think of Luis Alvarez, the physicist. He's a guy I admire for his gutsiness and sense of humor, and I think, "If Alvarez was on this commission, he would do it, and that's good enough for me."

There are stories of physicists—great heroes—who have gotten information one, two, three—just like that—

where everybody else is trying to do it in a complicated way. For example, after ultraviolet rays and X-rays had been discovered, there was a new type, called N-rays, discovered by André Blondel, in France. It was hard to detect the N-rays: other scientists had difficulty repeating Blondel's experiments, so someone asked the great American physicist R. W. Wood to go to Blondel's laboratory.

Blondel gave a public lecture and demonstration. N-rays were bent by aluminum, so he had all kinds of lenses lined up, followed by a big disk with an aluminum prism in the middle. As the aluminum prism slowly turned, the N-rays came up this way and bent that way, and Blondel's assistant reported their intensity—different numbers for different angles.

N-rays were affected by light, so Blondel turned out the lights to make his readings more sensitive. His assistant continued to report their intensity.

When the lights came back on, there's R. W. Wood in the front row, holding the prism high in the air, balanced on the tips of his fingers, for all to see! So that was the end of the N-ray.

I think, "Exactly! I've got to get a sample of the rubber." I call Bill Graham.

It's impossible to get: it's kept somewhere down at Kennedy. But then Graham remembers that the model of the field joint we're going to use in our meeting tomorrow has two samples of the rubber in it. He says, "We could meet in my office before the meeting and see if we can get the rubber out."

The next morning I get up early and go out in front of my hotel. It's eight in the morning and it's snowing. I find a taxi and say to the driver, "I'd like to go to a hardware store."

"A hardware store, sir?"

"Yeah. I gotta get some tools."

"Sir, there's no hardware stores around here; the Capitol is over there, the White House is over there—wait a minute: I think I remember passing one the other day."

He found the hardware store, and it turned out it didn't open till 8:30—it was about 8:15—so I waited outside, in my suitcoat and tie, a costume I had assumed since I came to Washington in order to move among the natives without being too conspicuous.

The suitcoats that the natives wear inside their buildings (which are well heated) are sufficient for walking from one building to another—or from a building to a taxi if the buildings are too far apart. (All the taxis are heated.) But the natives seem to have a strange fear of the cold: they put overcoats on top of their suitcoats if they wish to step outside. I hadn't bought an overcoat yet, so I was still rather conspicuous standing outside the hardware store in the snow.

At 8:30 I went in and bought a couple of screwdrivers, some pliers, and the smallest C-clamp I could find. Then I went to NASA.

On the way to Graham's office, I thought maybe the clamp was too big. I didn't have much time, so I ran down to the medical department of NASA. (I knew where it was, because I had been going there for blood tests ordered by my cardiologist, who was trying to treat me by telephone.) I asked for a medical clamp like they put on tubes.

They didn't have any. But the guy says, "Well, let's see if your C-clamp fits inside a glass!" It fitted very easily.

I went up to Graham's office.

The rubber came out of the model easily with just a pair of pliers. So there I was with the rubber sample in my hand. Although I knew it would be more dramatic and honest to do the experiment for the first time in the public meeting, I did something that I'm a little bit ashamed of. I cheated. I couldn't resist. I tried it. So, following the example of having a closed meeting before an open meet-

FIGURE 15. The field-joint model from which Feynman got the O-ring sample.

ing, I discovered it worked before I did it in the open meeting. Then I put the rubber back into the model so Graham could take it to the meeting.

I go to the meeting, all ready, with pliers in one pocket and a C-clamp in the other. I sit down next to General Kutyna.

At the previous meeting, there was ice water for everybody. This time, there's no ice water. I get up and go over to somebody who looks like he's in charge, and I say, "I'd like a glass of ice water, please."

He says, "Certainly! Certainly!"

Five minutes later, the guards close the doors, the meeting starts, and I haven't got my ice water.

I gesture over to the guy I just talked to. He comes over and says, "Don't worry, it's coming!"

The meeting is going along, and now Mr. Mulloy begins to tell us about the seals. (Apparently, NASA wants to tell us about the seals before Mr. Cook does.) The model starts to go around, and each commissioner looks at it a little bit.

Meanwhile, no ice water!

Mr. Mulloy explains how the seals are supposed to work—in the usual NASA way: he uses funny words and acronyms, and it's hard for anybody else to understand.

In order to set things up while I'm waiting for the ice water, I start out: "During a launch, there are vibrations which cause the rocket joints to move a little bit—is that correct?"

"That is correct, sir."

"And inside the joints, these so-called O-rings are supposed to expand to make a seal—is that right?"

"Yes, sir. In static conditions they should be in direct contact with the tang and clevis* and squeezed twenty-thousandths of an inch."

"Why don't we take the O-rings out?"

"Because then you would have hot gas expanding through the joint . . ."

"Now, in order for the seal to work correctly, the O-rings must be made of rubber—not something like lead, which, when you squash it, it stays."

"Yes, sir."

"Now, if the O-ring weren't resilient for a second or two, would that be enough to be a very dangerous situation?"

"Yes, sir."

That led us right up to the question of cold temperature and the resilience of the rubber. I wanted to prove that Mr. Mulloy must have known that temperature had an effect, although—according to Mr. MacDonald—he claimed that the evidence was "incomplete." But still, no ice water! So I had to stop, and somebody else started asking questions.

*The tang is the male part of the joint; the clevis is the female part (see Figure 13).

The model comes around to General Kutyna, and then to me. The clamp and pliers come out of my pocket, I take the model apart, I've got the O-ring pieces in my hand, but I still haven't got any ice water! I turn around again and signal the guy I've been bothering about it, and he signals back, "Don't worry, you'll get it!"

Pretty soon I see a young woman, way down in front, bringing in a tray with glasses on it. She gives a glass of ice water to Mr. Rogers, she gives a glass of ice water to Mr. Armstrong, she works her way back and forth along the rows of the dais, giving ice water to everybody! The poor woman had gotten everything together—jug, glasses, ice, tray, the whole thing—so that everybody could have ice water.

So finally, when I get my ice water, I don't drink it! I squeeze the rubber in the C-clamp, and put them in the glass of ice water.

After a few minutes, I'm ready to show the results of my little experiment. I reach for the little button that activates my microphone.

General Kutyna, who's caught on to what I'm doing, quickly leans over to me and says, "Co-pilot to pilot: not now."

Pretty soon, I'm reaching for my microphone again.

"Not now!" He points in our briefing book—with all the charts and slides Mr. Mulloy is going through—and says, "When he comes to this slide, here, that's the right time to do it."

Finally Mr. Mulloy comes to the place, I press the button for my microphone, and I say, "I took this rubber from the model and put it in a clamp in ice water for a while."

I take the clamp out, hold it up in the air, and loosen it as I talk: "I discovered that when you undo the clamp, the rubber doesn't spring back. In other words, for more than a few seconds, there is no resilience in this particular material when it is at a temperature of 32 degrees. I believe that

FIGURE 15A. _The O-ring ice-water demonstration._ (© MARILYNN K. YEE, NYT PICTURES.)

has some significance for our problem."

Before Mr. Mulloy could say anything, Mr. Rogers says, "That is a matter we will consider, of course, at length in the session that we will hold on the weather, and I think it is an important point which I'm sure Mr. Mulloy acknowledges and will comment on in a further session."

During the lunch break, reporters came up to me and asked questions like, "Were you talking about the O-ring, or the putty?" and "Would you explain to us what an O-ring is, exactly?" So I was rather depressed that I wasn't able to make my point. But that night, all the news shows caught on to the significance of the experiment, and the next day, the newspaper articles explained everything perfectly.

Check Six!

MY cousin Frances educated me about the press. She had been the AP White House correspondent during the Nixon and Ford administrations, and was now working for CNN. Frances would tell me stories of guys running out back doors because they're afraid of the press. From her I got the idea that the press isn't doing anything evil; the reporters are simply trying to help people know what's going on, and it doesn't do any harm to be courteous to them.

I found out that they're really quite friendly, if you give them a chance. So I wasn't afraid of the press, and I would always answer their questions.

Reporters would explain to me that I could say, "Not for attribution." But I didn't want any hocus-pocus. I didn't want it to sound like I'm leaking something. So whenever I talked to the press, I was straight. As a result of this, my name was in the newspaper every day, all over the place!

It seemed like I was always the one answering the reporters' questions. Often the rest of the commissioners would be anxious to go off to lunch, and I'd still be there, answering questions. But I figured, "What's the point of having a public meeting if you run away when they ask you what a word meant?"

When we'd finally get to our lunch, Mr. Rogers would remind us to be careful not to talk to the press. I would say something like, "Well, I was just telling them about the O-rings."

He would say, "That's okay. You've been doing all right, Dr. Feynman; I have no problem with that." So I never did figure out, exactly, what he meant by "not talking to the press."

Being on the commission was rather tense work, so I enjoyed having dinner once in a while with Frances and Chuck, my sister's son, who was working for the *Washington Post*. Because Mr. Rogers kept talking about leaks, we made sure we never said a *word* about anything I was doing. If CNN needed to find out something from me, they'd have to send a different reporter. The same went for the *Post*.

I told Mr. Rogers about my relatives working for the press: "We've agreed not to talk about my work. Do you think there's any problem?"

He smiled and said, "It's perfectly all right. I have a cousin in the press, too. There's no problem at all."

On Wednesday the commission had nothing to do, so General Kutyna invited me over to the Pentagon to educate me on the relationship between the air force and NASA.

It was the first time I had ever been in the Pentagon. There were all these guys in uniform who would take orders—not like in civilian life. He says to one of them, "I'd like to use the briefing room . . ."

"Yes, sir!"

". . . and we'll need to see slides number such-and-such and so-and-so."

"Yes, sir! Yes, sir!"

We've got all these guys working for us while General Kutyna gives me a big presentation in this special briefing room. The slides are shown from the back on a transparent wall. It was really fancy.

General Kutyna would say things like, "Senator So-and-so is in NASA's pocket," and I would say, half-joking,

"Don't give me these side remarks, General; you're filling my head! But don't worry, I'll forget it all." I wanted to be naive: I'd find out what happened to the shuttle first; I'd worry about the big political pressures later.

Somewhere in his presentation, General Kutyna observed that everybody on the commission has some weakness because of their connections: he, having worked very closely with NASA personnel in his former position as Air Force Space Shuttle Program manager, finds it difficult, if not impossible, to drive home some of the tougher questions on NASA management. Sally Ride still has a job with NASA, so she can't just say everything she wants. Mr. Covert had worked on the engines, and had been a consultant to NASA, and so on.

I said, "I'm associated with Caltech, but I don't consider that a weakness!"

"Well," he says, "that's right. You're invincible—as far as we can see. But in the air force we have a rule: check six."

He explained, "A guy is flying along, looking in all directions, and feeling very safe. Another guy flies up behind him (at 'six o'clock'—'twelve o'clock' is directly in front), and shoots. Most airplanes are shot down that way. Thinking that you're safe is very dangerous! Somewhere, there's a weakness you've got to find. You must always check six o'clock."

An underling comes in. There's some mumbling about somebody else needing the briefing room now. General Kutyna says, "Tell them I'll be finished in ten minutes."

"Yes, sir!"

Finally, we go out. There, in the hall, are TEN GENERALS waiting to use the room—and I had been sitting in there, getting this personal briefing. I felt great.

For the rest of the day, I wrote a letter home. I began to worry about "check six" when I described Mr. Rogers' reaction to my visiting Frances and Chuck. I wrote,

... I was pleased by Rogers' reaction, but now as I write this I have second thoughts. It was too easy—after he explicitly talked about the importance of no leaks etc. at earlier meetings. Am I being set up? (SEE, DARLING, <u>WASHINGTON PARANOIA IS SETTING IN</u>.) ... I think it is possible that there are things in this somebody might be trying to keep me from finding out and might try to discredit me if I get too close. ... So, reluctantly, I will have to not visit Frances and Chuck any more. Well, I'll ask Fran first if that is *too* paranoid. Rogers seemed so agreeable and reassuring. It was so easy, yet I am probably a thorn in his side. ...

Tomorrow at 6:15 am we go by special airplane (two planes) to Kennedy Space Center to be "briefed." No doubt we shall wander about, being shown everything—gee whiz—but no time to get into technical details with anybody. Well, it won't work. If I am not satisfied by Friday, I will stay over Sat & Sun, or if they don't work then, Monday & Tuesday. I am determined to do the job of finding out what happened—let the chips fall!

My guess is that I will be allowed to do this, overwhelmed with data and details . . . , so they have time to soften up dangerous witnesses etc. But it won't work because (1) I do technical information exchange and understanding much faster than they imagine, and (2) I already smell certain rats that I will not forget, because I just love the smell of rats, for it is the spoor of exciting adventure.

I feel like a bull in a china shop. The best thing is to put the bull out to work on the plow. A better metaphor will be an ox in a china shop, because the china is the bull, of course.*

So, much as I would rather be home and doing something else, I am having a wonderful time.

Love,
Richard

The press was reporting rumors that NASA was under great political pressure to launch the shuttle, and there were various theories as to where the pressure was coming

*The thing Feynman was going to break up was the baloney (the "bull——") about how good everything was at NASA.

from. It was a great big world of mystery to me, with tremendous forces. I would investigate it, all right, and if I
protected myself, nothing would happen. But I hadda
watch out.

FINALLY, early on Thursday morning, we get to Florida. The original idea was that we would go around the Kennedy Space Center at Cape Canaveral and see everything on a guided tour. But because information was coming out in the newspapers so fast, we had a public meeting first.

First, we saw some detailed pictures of the smoke coming out of the shuttle while it was still on the launch pad. There are cameras all over the place watching the launch—something like a hundred of them. Where the smoke came out, there were two cameras looking straight at it—but both failed, curiously. Nevertheless, from other cameras we could see four or five puffs of black smoke coming out from a field joint. This smoke was not burning material; it was simply carbon and mucky stuff that was pushed out because of pressure inside the rocket.

The puffs stopped after a few seconds: the seal got plugged up somehow, temporarily, only to break open again a minute later.

There was some discussion about how much matter came out in the smoke. The puffs of smoke were about six feet long, and a few feet thick. The amount of matter depends on how fine the particles are, and there could always be a big piece of glop inside the smoke cloud, so it's hard to judge. And because the pictures were taken from

Gumshoes

FIGURE 16. *Detailed picture, taken from the launch pad, of the "smoke." (© NASA.)*

the side, it was possible there was more smoke farther around the rocket.

To establish a minimum, I assumed a particle size that would produce as much smoke as possible out of a given amount of material. It came out surprisingly small—approximately one cubic inch: if you have a cubic inch of stuff, you can get that much smoke.

We asked for pictures from other launches. We found out later that there had never been any puffs of smoke on any previous flights.

We also heard about the low temperatures before the launch from a man named Charlie Stevenson, who was in charge of the ice crew. He said the temperature had gone down to 22 degrees during the night, but his crew got readings as low as 8 degrees at some places on the launch pad, and they couldn't understand why.

During the lunch break, a reporter from a local TV station asked me what I thought about the low temperature readings. I said it seemed to me that the liquid hydrogen and oxygen had chilled the 22-degree air even further as it flowed down the big fuel tank onto the rocket booster. For some reason, the reporter thought I had just told him some important, secret information, so he didn't use my name in his report that evening. Instead, he said, "This explanation comes from a Nobel Prize winner, so it must be right."

In the afternoon, the telemetering people gave us all kinds of information on the last moments of the shuttle. Hundreds of things had been measured, all of which indicated that everything was working as well as it could under the circumstances: the pressure in the hydrogen tank suddenly fell a few seconds after the flame had been observed; the gyros which steer the shuttle were working perfectly until one had to work harder than the other because there were side forces from the flame shooting out of the side of the booster rocket; the main engines even shut themselves down when the hydrogen tank exploded, because there was a pressure drop in the fuel lines.

That meeting lasted until 7:30 in the evening, so we postponed the tour until Friday and went straight to a dinner set up by Mr. Rogers.

At the dinner I happened to be seated next to Al Keel, who had joined the commission on Monday as its executive officer to help Mr. Rogers organize and run our work. He came to us from the White House—from something called the OMB*—and had a good reputation for doing a fine job at this and that. Mr. Rogers kept saying how lucky we were to get somebody with such high qualifications.

One thing that impressed me, though, was that Dr. Keel had a Ph.D. in aerospace, and had done some post-doc work at Berkeley. When he introduced himself on Monday, he joked that the last "honest work" he had done for a living was some aerodynamics work for the shuttle program ten or twelve years ago. So I felt very comfortable with him.

Well, I haven't been talking to Dr. Keel for more than five minutes, when he tells me he's never been so insulted in his life, that he didn't take this job to be so insulted, and that he doesn't want to talk to me anymore!

Now, I have a way of not remembering things when I do something dumb or annoying to people, so I forget what I said that put him out. Whatever it was, I thought I was joking, so I was very surprised by his reaction. I had undoubtedly said some boorish, brash, damn-fool thing, which I therefore can't remember!

Then there was a rather tense period of five or ten minutes, with me apologizing and trying to get a conversation going again. We finally got to talking again, somewhat. We were not big friends, but at least there was peace.

On Friday morning, we had another public meeting, this time to hear people from Thiokol and NASA talk about

---

*The Office of Management and Budget.

the night before the launch. Everything came out so slowly: the witness doesn't really want to tell you everything, so you have to get the answers out by asking exactly the right questions.

Other guys on the commission were completely awake—Mr. Sutter, for instance. "Exactly what were your quality criteria for acceptance under such-and-such and so-and-so?"—he'd ask specific questions like that, and it would turn out they didn't have any such criteria. Mr. Covert and Mr. Walker were the same way. Everybody was asking good questions, but I was fogged out most of the time, feeling a little bit behind.

Then this business of Thiokol changing its position came up. Mr. Rogers and Dr. Ride were asking two Thiokol managers, Mr. Mason and Mr. Lund, how many people were against the launch, even at the last moment.

"We didn't poll everyone," says Mr. Mason.

"Was there a substantial number against the launch, or just one or two?"

"There were, I would say, probably five or six in engineering who at that point would have said it is not as conservative to go with that temperature, and we don't know. The issue was we didn't know for sure that it would work."

"So it was evenly divided?"

"That's a very estimated number."

It struck me that the Thiokol managers were waffling. But I only knew how to ask simpleminded questions. So I said, "Could you tell me, sirs, the names of your four best seals experts, in order of ability?"

"Roger Boisjoly and Arnie Thompson are one and two. Then there's Jack Kapp and, uh . . . Jerry Burns."

I turned to Mr. Boisjoly, who was right there, at the meeting. "Mr. Boisjoly, were you in agreement that it was okay to fly?"

He says, "No, I was not."

I ask Mr. Thompson, who was also there.

"No, I was not."

I say, "Mr. Kapp?"

Mr. Lund says, "He is not here. I talked to him after the meeting, and he said, 'I would have made that decision, given the information we had.'"

"And the fourth man?"

"Jerry Burns. I don't know what his position was."

"So," I said, "of the four, we have one 'don't know,' one 'very likely yes,' and the two who were mentioned right away as being the *best* seal experts, *both said no.*" So this "evenly split" stuff was a lot of crap. The guys who knew the *most* about the seals—what were *they* saying?

Late in the afternoon, we were shown around the Kennedy Space Center. It was interesting; it wasn't as bad as I had predicted. The other commissioners asked a lot of important questions. We didn't have time to see the booster-rocket assembly, but near the end we were going to see the wreckage that had been recovered so far. I was pretty tired of this group stuff, so I excused myself from the rest of the tour.

I ran down to Charlie Stevenson's place to see more pictures of the launch. I also found out more about the unusually low temperature readings. The guys were very cooperative, and wanted me to work with them. I had been waiting for *ten days* to run around in one of these places, and here I was, *at last!*

At dinner that night, I said to Mr. Rogers, "I was thinking of staying here over the weekend."

"Well, Dr. Feynman," he said, "I'd prefer you come back to Washington with us tonight. But of course, you're free to do whatever you want."

"Well, then," I said, "I'll stay."

On Saturday I talked to the guy who had actually taken the temperature readings the morning of the launch—a nice fella named B. K. Davis. Next to each temperature he had written the exact time he had measured it, and then took a picture of it. You could see large gaps between the times as he climbed up and down the big launch tower. He measured the temperature of the air, the rocket, the ground, the ice, and even a puddle of slush with antifreeze in it. He did a very complete job.

NASA had a theoretical calculation of how the temperatures should vary around the launch pad: they should have been more uniform, and higher. Somebody thought that heat radiating to the clear sky had something to do with it. But then someone else noticed that BK's reading for the slush was much lower than the photograph indicated: at 8 degrees, the slush—even with antifreeze in it—should have been frozen solid.

Then we looked at the device the ice crew used for measuring the temperatures. I got the instruction manual out, and found that you're supposed to put the instrument out in the environment for at least 20 minutes before using it. Mr. Davis said he had taken it out of the box—at 70 degrees—and began making measurements right away. Therefore we had to find out whether the errors were reproducible. In other words, could the circumstances be duplicated?

On Monday I called up the company that made the device, and talked to one of their technical guys: "Hi, my name is Dick Feynman," I said. "I'm on the commission investigating the *Challenger* accident, and I have some questions about your infrared scanning gun . . ."

"May I call you right back?" he says.

"Sure."

After a little while he calls me back: "I'm sorry, but it's proprietary information. I can't discuss it with you."

By this time I realized what the real difficulty was: the company was *scared green* that we were going to blame the accident on their instrument. I said, "Sir, your scanning gun has nothing to do with the accident. It was used by the people here in a way that's contrary to the procedures in your instruction manual, and I'm trying to figure out if we can reproduce the errors and determine what the temperatures really were that morning. To do this, I need to know more about your instrument."

The guy finally came around, and became quite cooperative. With his help, I advised the ice-crew guys on an experiment. They cooled a room down to about 40 degrees, and put a big block of ice in it—with ice, you can be sure the surface temperature is 32 degrees. Then they brought in the scanning gun from a room which was 70 degrees inside, and made measurements of the ice block every 30 seconds. They were able to measure how far off the instrument was as a function of time.

Mr. Davis had written his measurements so carefully that it was very easy to fix all the numbers. And then, remarkably, the recalculated temperatures were close to what was expected according to the theoretical model. It looked very sensible.

The next time I talked to a reporter, I straightened everything out about the temperatures, and informed him that the earlier theory expounded by the Nobel Prize winner was wrong.

I wrote a report for the other commissioners on the temperature problem, and sent it to Dr. Keel.

Then I investigated something we were looking into as a possible contributing cause of the accident: when the booster rockets hit the ocean, they became out of round a little bit from the impact. At Kennedy they're taken apart, and the sections—four for each rocket—are sent by rail to Thiokol in Utah, where they are packed with new propellant. Then they're put back on a train to Florida. During transport, the sections (which are hauled on their side) get squashed a little bit—the softish propellant is very heavy. The total amount of squashing is only a fraction of an inch, but when you put the rocket sections back together, a small gap is enough to let hot gases through: the O-rings are only a quarter of an inch thick, and compressed only two-hundredths of an inch!

I thought I'd do some calculations. NASA gave me all the numbers on how far out of round the sections can get, so I tried to figure out how much the resulting squeeze was, and where it was located—maybe the minimum squeeze was where the leak occurred. The numbers were measurements taken along three diameters, every 60 degrees. But three matching diameters won't guarantee that things will fit; six diameters, or any other number of diameters, won't do, either.

For example, you can make a figure something like a triangle with rounded corners, in which three diameters, 60 degrees apart, have the same length.

I remembered seeing such a trick at a museum when I was a kid. There was a gear rack that moved back and forth perfectly smoothly, while underneath it were some noncircular, funny-looking, crazy-shaped gears turning on shafts that wobbled. It looked impossible, but the reason it worked was that the gears were shapes whose diameters were always the same.

So the numbers NASA gave me were useless.

During that weekend, just as I had predicted in my letter home, I kept getting notes from the commission headquarters in Washington: "Check the temperature readings, check the pictures, check this, check that . . ."—there was quite a list. But as the instructions came in, I had done most of them already.

One note had to do with a mysterious piece of paper. Someone at Kennedy had reportedly written "Let's go for it" while assembling one of the solid booster rockets. Such language appeared to show a certain recklessness. My mission: find that piece of paper.

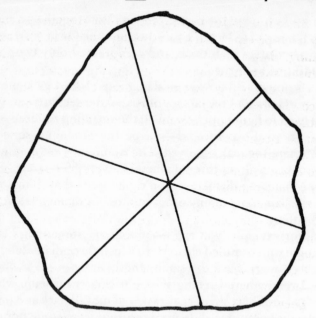

*FIGURE 17. This figure has all its diameters the same
length—yet it is obviously not round!*

Well, by this time I understood how much paper there
was in NASA. I was sure it was a trick to make me get lost,
so I did nothing about it.

Instead, I pursued something surreptitiously.

It was rumored that the reason NASA tried to make the
shuttle fly on January 28th, in spite of the cold, was that the
president was going to give his State of the Union address
that night. According to the theory, the White House had
it all cooked up so that during the State of the Union address, the teacher, Mrs. McAuliffe, would talk to the president and Congress from space. It was gonna be great: the
president would say, "Hello! How are you doing?" And she
would say, "Fine"—something very dramatic.

Since it sounded logical, I began by supposing it was very likely possible. But was there any evidence? This kind of thing I didn't know how to investigate. I could only think of this: it's very hard to get through to the president; I also can't just call up an astronaut and talk to her—if she's in space. Therefore, switching the signals down from the shuttle over to the president while he's talking to Congress must be a complicated business.

To find out whether anybody had set up to do that, I went down to the lowest levels and asked guys at the bottom some technical questions.

They showed me the antennas, they told me about the frequencies, they showed me the big radio system and the computer system; they showed me all the ways they did things.

I said, "If you had to send a transmission somewhere else—to Marshall, say—how would you do it?"

They said, "Oh, we're just a relay station. Everything is automatically sent over to Houston, and they switch everything out from there. We don't do any switching here."

So I didn't find any evidence—at least at Kennedy. But the guys there were so nice to me, and everything was so pleasant, that I feel bad. I don't like to cheat people. It was a little sneaky, what I was doing. Nevertheless, I thought I'd better do the same thing when I got to Houston.

On Monday, Mr. Hotz came down to Florida to work with me. (He told me later that he had been sent down with specific instructions to see what I was doing, and to keep me from "going wild.") Mr. Hotz brought a list of things to look into: "There are a lot of things on this list," he said, "so I'd be happy to split the work with you." Some things he said he could do more easily, and the rest of the things I had already done—except for that piece of paper which said "Let's go for it." Mr. Hotz hinted around that it might have come from the diary of someone in the booster-rocket assembly. That wasn't enough of a clue for me; I just wasn't gonna do it. Instead, I went to see a Mr. Lamberth, who had said he wanted to talk to me.

Mr. Lamberth was way up in the works, a big cheese in charge of assembling the solid-rocket boosters. He wanted to tell me about some problems he had. "The workers used to have much better discipline," he explained, "but nowadays they're not like they used to be." He gave me a couple of examples.

The first incident had to do with taking the booster rockets apart after they had been recovered from the sea. The rocket sections are held together by 180 pins—each about an inch and a half in diameter and two inches long—all the way around.

There was some kind of procedure for taking sections apart, in which the workers were supposed to pull the rocket up a certain distance. They had gotten to paying attention only to the amount of *force* they were applying—about 11,000 pounds. That was a better method, from a physical standpoint, because the idea is to take the load off the pins.

One time the force gauge wasn't working right. The workers kept putting more force on, wondering why they weren't reaching 11,000 pounds, when all of a sudden one of the pins broke.

Mr. Lamberth reprimanded the workers for not following procedures. It reminded me of when I tried to make things work better at my aunt's hotel: your method is better than the regular way, but then you have a little accident . . .*

The second story Mr. Lamberth told me had to do with putting the rocket sections together. The regular procedure was to stack one section on top of the other and match the upper section to the lower one.

*The reference is to Feynman's method of slicing string beans, recounted in *Surely You're Joking, Mr. Feynman!*

If a section needed to be reshaped a little bit, the procedure was to first pick up the section with a crane and let it hang sideways a few days. It's rather simpleminded.

If they couldn't make a section round enough by the hanging method, there was another procedure: use the "rounding machine"—a rod with a hydraulic press on one end and a nut on the other—and increase the pressure.

Mr. Lamberth told me the pressure shouldn't exceed 1200 pounds per square inch (psi). One time, a section wasn't round enough at 1200 psi, so the workers took a wrench and began turning the nut on the other end. When they finally got the section round enough, the pressure was up to 1350. "This is another example of the lack of discipline among the workers," Mr. Lamberth said.

I had wanted to talk with the assembly workers anyway (I love that kind of thing), so I arranged to see them the next day at 2:30 in the afternoon.

At 2:30 I walk into this room, and there's a long table with thirty or forty people—they're all sitting there with morose faces, very serious, ready to talk to The Commissioner.

I was terrified. I hadn't realized my terrible power. I could see they were worried. They must have been told I was investigating the errors they had made!

So right away I said, "I had nothin' to do, so I thought I'd come over and talk to the guys who put the rockets together. I didn't want everybody to stop working just 'cause I wanna find out something for my own curiosity; I only wanted to talk with the workers . . ."

Most of the people got up and left. Six or seven guys stayed—the crew who actually put the rocket sections together, their foreman, and some boss who was higher up in the system.

Well, these guys were still a little bit scared. They didn't really want to open up. The first thing I think to say is, "I have a question: when you measure the three diameters and all the diameters match, do the sections really fit together? It seems to me that you could have some bumps on one side and some flat areas directly across, so the three diameters would match, but the sections wouldn't fit."

"Yes, yes!" they say. "We get bumps like that. We call them nipples."

The only woman there said, "It's got nothing to do with me!"—and everybody laughed.

"We get nipples all the time," they continued. "We've been tryin' to tell the supervisor about it, but we never get anywhere!"

We were talking details, and that works wonders. I would ask questions based on what could happen theoretically, but to them it looked like I was a regular guy who knew about their technical problems. They loosened up very rapidly, and told me all kinds of ideas they had to improve things.

For example, when they use the rounding machine, they have to put a rod through holes exactly opposite each other. There are 180 holes, so they have to make sure the other end of the rod goes through the hole 90 holes away. Now, it turns out you have to climb up into an awkward place to count the holes. It's very slow and very difficult.

They thought it would be very helpful if there were four paint marks, 90 degrees apart, put on at the factory. That way, they would never have to count more than 22 holes to the nearest mark. For example, if they put the rod through a hole which is 9 holes clockwise from a paint mark, then the other end of the rod would go through the hole which is 9 holes clockwise from the opposite mark.

The foreman, Mr. Fichtel, said he wrote a memo with this suggestion to his superiors two years ago, but nothing had happened yet. When he asked why, he was told the suggestion was too expensive.

"Too expensive to paint *four little lines?*" I said in disbelief.

They all laughed. "It's not the paint; it's the paperwork," Mr. Fichtel said. "They would have to revise all the manuals."

The assembly workers had other observations and suggestions. They were concerned that if two rocket sections scrape as they're being put together, metal filings could get into the rubber seals and damage them. They even had some suggestions for redesigning the seal. Those suggestions weren't very good, but the point is, the workers were *thinking!* I got the impression that they were *not* undisciplined; they were very interested in what they were doing, but they weren't being given much encouragement. Nobody was paying much attention to them. It was remarkable that their morale was as high as it was under the circumstances.

Then the workers began to talk to the boss who had stayed. "We're disappointed by something," one of them said. "When the commission was going to see the booster-rocket assembly, the demonstration was going to be done by the managers. Why wouldn't you let *us* do it?"

"We were afraid you'd be frightened by the commissioners and you wouldn't want to do it."

"No, no," said the workmen. "We think we do a good job, and we wanted to show what we do."

After that meeting, the boss took me to the cafeteria. As we were eating—the workmen weren't with us anymore—he said, "I was surprised they were so concerned about that."

Later, I talked to Mr. Fichtel about this incident of increasing the pressure past 1200. He showed me the notes he made as he went along: they weren't the formal papers that are stamped; they were part of an informal but carefully written diary.

I said, "I hear the pressure got up to 1350."

"Yes," he said, "we had tightened the nut at the other end."

"Was that the regular procedure?"

"Oh, yes," he said, "it's in the book."

He opens up the manual and shows me the procedure. It says, "Build up the pressure on the hydraulic jack. If this is insufficient to obtain desired roundness, then very carefully tighten nut on other end to get to the desired roundness"—it said so in black and white! It didn't say that tightening the nut would increase the pressure past 1200 psi; the people who wrote the manual probably weren't quite aware of that.

Mr. Fichtel had written in his diary, "We very carefully tightened the nut"—exactly the same language as the instructions.

I said, "Mr. Lamberth told me he admonished you about going above 1200."

"He never admonished me about that—why should he?"

We figured out what probably happened. Mr. Lamberth's admonishment went down through the levels until somebody in middle management realized that Mr. Fichtel had gone by the book, and that the error was in the manual. But instead of telling Mr. Lamberth about the error, they simply threw away the admonishment, and just kept quiet.

Over lunch, Mr. Fichtel told me about the inspection procedures. "There's a sheet for each procedure, like this one for the rounding procedure," he said. "On it there are boxes for stamps—one from the supervisor, one from quality control, one from the manufacturer, and for the bigger jobs, one from NASA."

He continued, "We make the measurements, go through one course of rounding, and then make the measurements again. If they don't match well enough, we repeat the steps. Finally, when the diameter differences are small enough, we go for it."

I woke up. "What do you mean, 'go for it'?" I said. "It sounds sort of cavalier . . ."

"No, no," he says. "That's just the lingo we use when we mean that all the conditions are satisfied, and we're ready to move to the next phase of the operation."

"Do you ever write that down—that 'go for it'?"

"Yes, sometimes."

"Let's see if we can find a place where you wrote it."

Mr. Fichtel looked through his diary, and found an example. The expression was completely natural to him—it wasn't reckless or cavalier; it was just his way of speaking.

On Monday and Tuesday, while I was running around down at Kennedy, Mr. Rogers was in Washington appearing before a Senate committee. Congress was considering whether it should have its own investigation.

Senator Hollings, from South Carolina, was giving Mr. Rogers a hard time: "Secretary Rogers," he says, "I'm anxious that you have an adequate staff thayah. How many *investigators* does yo' commission have?"

Mr. Rogers says, "We don't have investigators in the police sense. We're reading documents, understanding what they mean, organizing hearings, talking to witnesses—that sort of thing. We'll have an adequate staff, I assure you."

"Well, that's the point," Senator Hollings says. "From my experience in investigating cases, I'd want four or five investigators steeped in science and space technology going around down there at Canaveral talking to everybody, eating lunch with them. You'd be amazed, if you eat in the restaurants around there for two or three weeks, what you'll find out. You can't just sit and read what's given to you."

"We're not just going to sit and read," Mr. Rogers says

defensively. "We've gotten a lot of people in a room and asked them questions all at the same time, rather than have a gumshoe walking around, talking to people one at a time."

"I understand," says Senator Hollings. "Yet I'm concerned about yo' product if you don't have some gumshoes. That's the trouble with presidential commissions; I've been on 'em: they go on what's *fed* to 'em, and they don't look behind it. Then we end up with investigative reporters, people writing books, and everything else. People are *still* investigating the Warren Commission Report around this town."*

Mr. Rogers calmly says, "I appreciate your comments, Senator. You'll be interested to know that one of our commission members—he's a Nobel laureate—is down there in Florida today, investigating in the way you'd like him to investigate."

(Mr. Rogers didn't know it, but I was actually eating lunch with some engineers when he said that!)

Senator Hollings says, "I'm not questioning the competence of the Nobel laureate; I've been reading with great interest what he said. There's no question about the competence of the commission itself. It's just that when you investigate a case, you need investigators. You have already brought to the public's attention a lot of very interesting facts, so I think you haven't been negligent in any fashion."

So I saved Mr. Rogers a little bit. He saw that he had an answer for Mr. Hollings by the *good luck* that I stayed in Florida anyway, against his wishes!

*Note for foreign readers: the Warren Report was issued in 1964 by the Warren Commission, headed by retired Supreme Court Chief Justice Earl Warren, which investigated the assassination of President John F. Kennedy.

ON Tuesday afternoon I flew back to Washington, and went to the next meeting of the commission, on Wednesday. It was another public meeting. A manager of the Thiokol Company named Mr. Lund was testifying. On the night before the launch, Mr. Mulloy had told him to put on his "management hat" instead of his "engineering hat," so he changed his opposition to launch and overruled his own engineers. I was asking him some harsh questions when suddenly I had this feeling of the Inquisition.

Mr. Rogers had pointed out to us that we ought to be careful with these people, whose careers depend on us. He said, "We have all the advantages: we're sitting up here; they're sitting down there. They have to answer our questions; we don't have to answer their questions." Suddenly, all this came back to me and I felt terrible, and I couldn't do it the next day. I went back to California for a few days, to recover.

While I was in Pasadena, I went over to JPL and met with Jerry Solomon and Meemong Lee. They were studying the flame which appeared a few seconds before the main fuel tank exploded, and were able to bring out all kinds of details. (JPL has good enhancers of TV pictures from all their experience with planetary missions.) Later, I took the enhancements over to

Fantastic Figures

Charlie Stevenson and his crew at Kennedy to expedite things.

Somewhere along the line, somebody from the staff brought me something to sign: it said that my expenses were so-and-so much, but they weren't—they were more. I said, "This is not the amount I actually spent."

The guy said, "I know that, sir; you're only allowed a maximum of $75 a day for the hotel and food."

"Then why did you guys set me up in a hotel which costs $80 or $90 a night, and then you give me only $75 a day?"

"Yes, I agree; it's too bad, but that's the way it goes!"

I thought of Mr. Rogers' offer to put me in a "good hotel." What did he mean by that—that it would cost me *more*?

If you're asked to contribute months of time and effort to the government (and you lose money you would have made consulting for a company), the government ought to appreciate it a little more than to be cheap about paying you back. I'm not trying to make money off the government, but I'm not wanting to *lose* money, either! I said, "I'm not going to sign this."

Mr. Rogers came over and promised he would straighten it out, so I signed the paper.

I really think Mr. Rogers tried to fix it, but he was unable to. I thought of fighting this one to the end, but then I realized it's impossible: if I had been paid for my actual expenses, then of course all the other commissioners would have to be paid, too. That would be all right, but it would also mean that this commission was the only commission to be paid its actual expenses—and pretty soon, word would get out.

They have a saying in New York: "You can't fight City Hall," meaning "It's impossible." But this time, it was a hell of a lot bigger than City Hall: the $75 a day rule is a law of the United States! It might have been fun to fight it to the end, but I guess I was tired—I'm not as young as I used to be—so I just gave up.

Somebody told me they heard commissioners make $1000 a day, but the truth is, our government doesn't even pay their costs.

At the beginning of March, about a month after the commission started, we finally split up into working groups: the Pre-Launch Activities group was headed by Mr. Acheson; Mr. Sutter was in charge of the Design, Development, and Production panel; General Kutyna was leader of the Accident Analysis group; and Dr. Ride was in charge of the Mission Planning and Operations group.

I spent most of my time in Kutyna's group. I was in Ride's group, too, but I ended up not doing very much for her.

General Kutyna's group went to Marshall Space Flight Center in Huntsville, Alabama, to do its work. The first thing that happened there was, a man named Ullian came in to tell us something. As range safety officer at Kennedy, Mr. Ullian had to decide whether to put destruct charges on the shuttle. (If a rocket goes out of control, the destruct charges enable it to be blown up into small bits. That's much less perilous than a rocket flying around loose, ready to explode when it hits the ground.)

Every unmanned rocket has these charges. Mr. Ullian told us that 5 out of 127 rockets that he looked at had failed—a rate of about 4 percent. He took that 4 percent and divided it by 4, because he assumed a manned flight would be safer than an unmanned one. He came out with about a 1 percent chance of failure, and that was enough to warrant the destruct charges.

But NASA told Mr. Ullian that the probability of failure was more like 1 in 10^5.

I tried to make sense out of that number. "Did you say 1 in 10^5?"

"That's right; 1 in 100,000."

"That means you could fly the shuttle *every day* for an average of *300 years* between accidents—every day, one flight, for 300 years—which is obviously crazy!"

"Yes, I know," said Mr. Ullian. "I moved my number up to 1 in 1000 to answer all of NASA's claims—that they were much more careful with manned flights, that the typical rocket isn't a valid comparison, et cetera—and put the destruct charges on anyway."

But then a new problem came up: the Jupiter probe, *Galileo,* was going to use a power supply that runs on heat generated by radioactivity. If the shuttle carrying *Galileo* failed, radioactivity could be spread over a large area. So the argument continued: NASA kept saying 1 in 100,000 and Mr. Ullian kept saying 1 in 1000, at best.

Mr. Ullian also told us about the problems he had in trying to talk to the man in charge, Mr. Kingsbury: he could get appointments with underlings, but he never could get through to Kingsbury and find out how NASA got its figure of 1 in 100,000. The details of the story I can't remember exactly, but I thought Mr. Ullian was doing everything sensibly.

Our panel supervised the tests that NASA was doing to discover the properties of the seals—how much pressure the putty could take, and so on—in order to find out exactly what had happened. General Kutyna didn't want to jump to conclusions, so we went over and over things, checking all the evidence and seeing how well everything fitted together.

There was an awful lot of detailed discussion about exactly what happened in the last few seconds of the flight, but I didn't pay much attention to any of it. It was as though a train had crashed because the track had a gap in it, and we were analyzing which cars broke apart first, which cars broke apart second, and why some car turned over on its side. I figured once the train goes off the track, it doesn't make any difference—it's done. I became bored.

So I made up a game for myself: "Imagine that something else had failed—the main engines, for instance—and we were making the same kind of intensive investigation as we are now: would we discover the same slipping safety criteria and lack of communication?"

I thought I would do my standard thing—find out from the engineers how the engine works, what all the dangers are, what problems they've had, and everything else—and then, when I'm all loaded up so I know what I'm talking about, I'd confront whoever was claiming the probability of failure was 1 in 100,000.

I asked to talk to a couple of engineers about the engines. The guy says, "Okay, I'll fix it up. Is nine tomorrow morning okay?"

This time there were three engineers, their boss, Mr. Lovingood, and a few assistants—about eight or nine people.

Everybody had big, thick notebooks, full of papers, all nicely organized. On the front they said:

REPORT ON MATERIAL GIVEN TO COMMISSIONER RICHARD P. FEYNMAN ON MARCH WA-WA,* 1986.

I said, "Geez! You guys must have worked hard all night!"

"No, it's not so much work; we just put in the regular papers that we use all the time."

I said, "I just wanted to talk to a few engineers. There are so many problems to work on, I can't expect you all to stay here and talk to me."

But this time, everybody stayed.

Mr. Lovingood got up and began to explain everything to me in the usual NASA way, with charts and graphs which matched the information in my big book—all with bullets, of course.

*Feynman's way of saying, "whatever it was."

I won't bother you with all the details, but I wanted to understand everything about the engine. So I kept asking my usual dumb-sounding questions.

After a while, Mr. Lovingood says, "Dr. Feynman, we've been going for two hours, now. There are 123 pages, and we've only covered 20 so far."

My first reaction was to say, "Well, it isn't really going to take such a long time. I'm always a little slow at the beginning; it takes me a while to catch on. We'll be able to go much faster near the end."

But then I had a second thought. I said, "In order to speed things up, I'll tell you what I'm doing, so you'll know where I'm aiming. I want to know whether there's the same lack of communication between the engineers and the management who are working on the engine as we found in the case of the booster rockets."

Mr. Lovingood says, "I don't think so. As a matter of fact, although I'm now a manager, I was trained as an engineer."

"All right," I said. "Here's a piece of paper each. Please write on your paper the answer to this question: what do you think is the probability that a flight would be uncompleted due to a failure in this engine?"

They write down their answers and hand in their papers. One guy wrote "99-$\frac{44}{100}$% pure" (copying the Ivory soap slogan), meaning about 1 in 200. Another guy wrote something very technical and highly quantitative in the standard statistical way, carefully defining everything, that I had to translate—which also meant about 1 in 200. The third guy wrote, simply, "1 in 300."

Mr. Lovingood's paper, however, said,

> Cannot quantify. Reliability is judged from:
> - past experience
> - quality control in manufacturing
> - engineering judgment

"Well," I said, "I've got four answers, and one of them weaseled." I turned to Mr. Lovingood: "I think you weaseled."

"I don't think I weaseled."

"You didn't tell me *what* your confidence was, sir; you told me *how* you determined it. What I want to know is: after you determined it, what *was* it?"

He says, "100 percent"—the engineers' jaws drop, my jaw drops; I look at him, everybody looks at him—"uh, uh, minus epsilon!"

So I say, "Well, yes; that's fine. Now, the only problem is, WHAT IS EPSILON?"

He says, "10^{-5}." It was the same number that Mr. Ullian had told us about: 1 in 100,000.

I showed Mr. Lovingood the other answers and said, "You'll be interested to know that there *is* a difference between engineers and management here—a factor of more than 300."

He says, "Sir, I'll be glad to send you the document that contains this estimate, so you can understand it."*

I said, "Thank you very much. Now, let's get back to the engine." So we continued and, just like I guessed, we went faster near the end. I had to understand how the engine worked—the precise shape of the turbine blades,

---

*Later, Mr. Lovingood sent me that report. It said things like "The probability of mission success is necessarily very close to 1.0"—does that mean it *is* close to 1.0, or it *ought to be* close to 1.0?—and "Historically, this high degree of mission success has given rise to a difference in philosophy between unmanned and manned space flight programs; i.e., numerical probability versus engineering judgment." As far as I can tell, "engineering judgment" means they're just going to make up numbers! The probability of an engine-blade failure was given as a universal constant, as if all the blades were exactly the same, under the same conditions. The whole paper was quantifying everything. Just about every nut and bolt was in there: "The chance that a HPHTP pipe will burst is 10^{-7}." You can't estimate things like that; a probability of 1 in 10,000,000 is almost impossible to estimate. It was clear that the numbers for each part of the engine were chosen so that when you add everything together you get 1 in 100,000.

exactly how they turned, and so on—so I could understand its problems.

After lunch, the engineers told me all the problems of the engines: blades cracking in the oxygen pump, blades cracking in the hydrogen pump, casings getting blisters and cracks, and so on. They looked for these things with periscopes and special instruments when the shuttle came down after each flight.

There was a problem called "subsynchronous whirl," in which the shaft gets bent into a slightly parabolic shape at high speed. The wear on the bearings was so terrible—all the noise and the vibration—that it seemed hopeless. But they had found a way to get rid of it. There were about a dozen very serious problems; about half of them were fixed.

Most airplanes are designed "from the bottom up," with parts that have already been extensively tested. The shuttle, however, was designed "from the top down"—to save time. But whenever a problem was discovered, a lot of redesigning was required in order to fix it.

Mr. Lovingood isn't saying much now, but different engineers, depending on which problem it is, are telling me all this stuff, just like I could have found out if I went down to the engineers at Thiokol. I gained a great deal of respect for them. They were all very straight, and everything was great. We went all the way down to the end of the book. We made it.

Then I said, "What about this high-frequency vibration where some engines get it and others don't?"*

There's a quick motion, and a little stack of papers

---

*I had heard about this from Bill Graham. He said that when he was first on the job as head of NASA, he was looking through some reports and noticed a little bullet: "•4,000 cycle vibration is within our data base." He thought that was a funny-looking phrase, so he began asking questions. When he got all the way through, he discovered it was a rather serious matter: some of the engines would vibrate so much that they couldn't be used. He used it as an example of how difficult it is to get information unless you go down and check on it yourself.

appears. It's all put together nicely; it fits nicely into my book. It's all about the 4000-cycle vibration!

Maybe I'm a little dull, but I tried my best not to accuse anybody of anything. I just let them show me what they showed me, and acted like I didn't see their trick. I'm not the kind of investigator you see on TV, who jumps up and accuses the corrupt organization of withholding information. But I was fully aware that they hadn't told me about the problem until I asked about it. I usually acted quite naive—which I was, for the most part.

At any rate, the engineers all leaped forward. They got all excited and began to describe the problem to me. I'm sure they were delighted, because technical people love to discuss technical problems with technical people who might have an opinion or a suggestion that could be useful. And of course, they were very anxious to cure it.

They kept referring to the problem by some complicated name—a "pressure-induced vorticity oscillatory wawa," or something.

I said, "Oh, you mean a whistle!"

"Yes," they said; "it exhibits the characteristics of a whistle."

They thought the whistle could be coming from a place where the gas rushed through a pipe at high speed and split into three smaller pipes—where there were two partitions. They explained how far they had gotten in figuring out the problem.

When I left the meeting, I had the definite impression that I had found the same game as with the seals: management reducing criteria and accepting more and more errors that weren't designed into the device, while the engineers are screaming from below, "HELP!" and "This is a RED ALERT!"

The next evening, on my way home in the airplane, I was having dinner. After I finished buttering my roll, I took the little piece of thin cardboard that the butter pat comes on, and bent it around in a U shape so there were two edges facing me. I held it up and started blowing on it, and pretty soon I got it to make a noise like a whistle.

Back in California, I got some more information on the shuttle engine and its probability of failure. I went to Rocketdyne and talked to engineers who were building the engines. I also talked to consultants for the engine. In fact one of them, Mr. Covert, was on the commission. I also found out that a Caltech professor had been a consultant for Rocketdyne. He was very friendly and informative, and told me about all the problems the engine had, and what he thought the probability of failure was.

I went to JPL and met a fellow who had just written a report for NASA on the methods used by the FAA* and the military to certify their gas turbine and rocket engines. We spent the whole day going back and forth over how to determine the probability of failure in a machine. I learned a lot of new names—like "Weibull," a particular mathematical distribution that makes a certain shape on a graph. He said that the original safety rules for the shuttle were very similar to those of the FAA, but that NASA had modified them as they began to get problems.

It turned out that NASA's Marshall Space Center in Huntsville designed the engine, Rocketdyne built them, Lockheed wrote the instructions, and NASA's Kennedy Space Center installed them! It may be a genius system of organization, but it was a complete fuzdazzle, as far as I was concerned. It got me terribly confused. I didn't know whether I was talking to the Marshall man, the Rocketdyne man, the Lockheed man, or the Kennedy man! So in the middle of all this, I got lost. In fact, all during this time—in March and April—I was running back and forth so much between California, Alabama, Houston, Florida, and Wash-

*Note for foreign readers: Federal Aviation Administration.

ington, D.C., that I often didn't know what day it was, or where I was.

After all this investigating on my own, I thought I'd write up a little report on the engine for the other commissioners. But when I looked at my notes on the testing schedules, there was some confusion: there would be talk about "engine #12" and how long "the engine" flew. But no engine ever was like that: it would be repaired all the time. After each flight, technicians would inspect the engines and see how many cracked blades there were on the rotor, how many splits there were in the casing, and so on. Then they'd repair "the engine" by putting on a new casing, a new rotor, or new bearings—they would replace lots of parts. So I would read that a particular engine had rotor #2009, which had run for 27 minutes in flight such-and-such, and casing #4091, which had run for 53 minutes in flights such-and-such and so-and-so. It was all mixed up.

When I finished my report, I wanted to check it. So the next time I was at Marshall, I said I wanted to talk to the engineers about a few very technical problems, just to check the details—I didn't need any management there.

This time, to my surprise, nobody came but the three engineers I had talked to before, and we straightened everything out.

When I was about to leave, one of them said, "You know that question you asked us last time—with the papers? We felt that was a loaded question. It wasn't fair."

I said, "Yes, you're quite right. It *was* a loaded question. I had an idea of what would happen."

The guy says, "I would like to revise my answer. I want to say that I cannot quantify it." (This guy was the one who had the most detailed answer before.)

I said, "That's fine. But do you agree that the chance of failure is 1 in 100,000?"

"Well, uh, no, I don't. I just don't want to answer."

Then one of the other guys says, "I said it was 1 in 300, and I still say it's 1 in 300, but I don't want to tell you how I got my number."

I said, "It's okay. You don't have to."

ALL during this time, I had the impression that somewhere along the line the whole commission would come together again so we could talk to each other about what we had found out.

In order to aid such a discussion, I thought I'd write little reports along the way: I wrote about my work with the ice crew (analyzing the pictures and the faulty temperature readings); I wrote about my conversations with Mr. Lamberth and the assembly workers; and I even wrote about the piece of paper that said "Let's go for it." All these little reports I sent to Al Keel, the executive officer, to give to the other commissioners.

Now, this particular adventure—investigating the lack of communication between the managers and the engineers who were working on the engine—I also wrote about, on my little IBM PC at home. I was kind of tired, so I didn't have the control I wanted—it wasn't written with the same care as my other reports. But since I was writing it only as a report to the other commissioners, I didn't change the language before I sent it on to Dr. Keel. I simply attached a note that said "I think the other commissioners would be interested in this, but you can do with it what you want—it's a little strong at the end."

He thanked me, and said he sent my report to everybody.

An Inflamed Appendix

Then I went to the Johnson Space Center, in Houston, to look into the avionics. Sally Ride's group was there, investigating safety matters in connection with the astronauts' experiences. Sally introduced me to the software engineers, and they gave me a tour of the training facilities for the astronauts.

It's really quite wonderful. There are different kinds of simulators with varying degrees of sophistication that the astronauts practice on. One of them is just like the real thing: you climb up, you get in; at the windows, computers are producing pictures. When the pilot moves the controls, the view out of the windows changes.

This particular simulator had the double purpose of teaching the astronauts and checking the computers. In the back of the crew area, there were trays full of cables running down through the cargo bay to somewhere in the back, where instruments simulated signals from the engines—pressures, fuel flow rates, and so on. (The cables were accessible because the technicians were checking for "cross talk"—interferences in the signals going back and forth.)

The shuttle itself is operated essentially by computer. Once it's lit up and starts to go, nobody inside does anything, because there's tremendous acceleration. When the shuttle reaches a certain altitude, the computers adjust the engine thrust down for a little while, and as the air thins out, the computers adjust the thrust up again. About a minute later, the two solid rocket boosters fall away, and a few minutes after that, the main fuel tank falls away; each operation is controlled by the computers. The shuttle gets into orbit automatically—the astronauts just sit in their seats.

The shuttle's computers don't have enough memory to hold all the programs for the whole flight. After the shuttle gets into orbit, the astronauts take out some tapes and load in the program for the next phase of the flight—

there are as many as six in all. Near the end of the flight, the astronauts load in the program for coming down.

The shuttle has four computers on board, all running the same programs. All four are normally in agreement. If one computer is out of agreement, the flight can still continue. If only two computers agree, the flight has to be curtailed and the shuttle brought back immediately.

For even more safety, there's a fifth computer—located away from the other four computers, with its wires going on different paths—which has only the program for going up and the program for coming down. (Both programs can barely fit into its memory.) If something happens to the other computers, this fifth computer can bring the shuttle back down. It's never had to be used.

The most dramatic thing is the landing. Once the astronauts know where they're supposed to land, they push one of three buttons—marked Edwards, White Sands, and Kennedy—which tells the computer where the shuttle's going to land. Then some small rockets slow the shuttle down a little, and get it into the atmosphere at just the right angle. That's the dangerous part, where all the tiles heat up.

During this time, the astronauts can't see anything, and everything's changing so fast that the descent has to be done automatically. At around 35,000 feet the shuttle slows down to less than the speed of sound, and the steering can be done manually, if necessary. But at 4000 feet something happens that is not done by the computer: the pilot pushes a button to lower the landing wheels.

I found that very odd—a kind of silliness having to do with the psychology of the pilots: they're heroes in the eyes of the public; everybody has the idea that they're steering the shuttle around, whereas the truth is they don't have to do anything until they push that button to lower the landing gear. They can't stand the idea that they really have nothing to do.

I thought it would be safer if the computer would lower the landing wheels, in case the astronauts were unconscious for some reason. The software engineers agreed, and added that putting down the landing wheels at the wrong time is very dangerous.

The engineers told me that ground control can send up the signal to lower the landing wheels, but this backup gave them some pause: what happens if the pilot is half-conscious, and thinks the wheels should go down at a certain time, and the controller on the ground knows it's the wrong time? It's much better to have the whole thing done by computer.

The pilots also used to control the brakes. But there was lots of trouble: if you braked too much at the beginning, you'd have no more brake-pad material left when you reached the end of the runway—and you're still moving! So the software engineers were asked to design a computer program to control the braking. At first the astronauts objected to the change, but now they're very delighted because the automatic braking works so well.

Although there's a lot of good software being written at Johnson, the computers on the shuttle are so obsolete that the manufacturers don't make them anymore. The memories in them are the old kind, made with little ferrite cores that have wires going through them. In the meantime we've developed much better hardware: the memory chips of today are much, much smaller; they have much greater capacity; and they're much more reliable. They have internal error-correcting codes that automatically keep the memory good. With today's computers we can design separate program modules so that changing the payload doesn't require so much program rewriting.

Because of the huge investment in the flight simulators and all the other hardware, to start all over again and replace the millions of lines of code that they've already built up would be very costly.

I learned how the software engineers developed the avionics for the shuttle. One group would design the software programs, in pieces. After that, the parts would be put together into huge programs, and tested by an independent group.

After both groups thought all the bugs had been worked out, they would have a simulation of an entire flight, in which every part of the shuttle system is tested. In such cases, they had a principle: this simulation is not just an exercise to check if the programs are all right; it is a *real flight*—if anything fails now, it's extremely serious, as if the astronauts were really on board and in trouble. Your reputation is on the line.

In the many years they had been doing this, they had had only six failures at the level of flight simulation, and not one in an actual flight.

So the computer people looked like they knew what they were doing: they knew the computer business was vital to the shuttle but potentially dangerous, and they were being extremely careful. They were writing programs that operate a very complex machine in an environment where conditions are changing drastically—programs which measure those changes, are flexible in their responses, and maintain high safety and accuracy. I would say that in some ways they were once in the forefront of how to ensure quality in robotic or interactive computer systems, but because of the obsolete hardware, it's no longer true today.

I didn't investigate the avionics as extensively as I did the engines, so I might have been getting a little bit of a sales talk, but I don't think so. The engineers and the managers communicated well with each other, and they were all very careful not to change their criteria for safety.

I told the software engineers I thought their system

and their attitude were very good.

One guy muttered something about higher-ups in NASA wanting to cut back on testing to save money: "They keep saying we always pass the tests, so what's the use of having so many?"

Before I left Houston, I continued my surreptitious investigation of the rumor that the White House had put pressure on NASA to launch the shuttle. Houston is the center of communication, so I went over to the telemetry people and asked about their switching system. I went through the same stuff as I did in Florida—and they were just as nice to me—but this time I found out that if they wanted to tie in the shuttle to the Congress, the White House, or to anywhere, they need a three-minute warning—not three months, not three days, not three hours—three minutes. Therefore they can do it whenever they want, and nothing has to be written down in advance. So that was a blind alley.

I talked to a *New York Times* reporter about this rumor one time. I asked him, "How do you find out if things like this are true?"

He says, "One of the things I thought to do was to go down and talk to the people who run the switching system. I tried that, but I wasn't able to come up with anything."

During the first half of April, General Kutyna's group received the final results of the tests NASA was making at Marshall. NASA included its own interpretations of the results, but we thought we should write everything over again in our own way. (The only exceptions were when a test didn't show anything.)

General Kutyna set up a whole system at Marshall for writing our group's report. It lasted about two days. Before we could get anywhere, we got a message from Mr. Rogers:

"Come back to Washington. You shouldn't do the writing down there."

So we went back to Washington, and General Kutyna gave me an office in the Pentagon. It was fine, but there was no secretary, so I couldn't work fast.

Bill Graham had always been very cooperative, so I called him up. He arranged for me to use a guy's office—the guy was out of town—and his secretary. She was very, very helpful: she could write up something as fast as I could say it, and then she'd revamp it, correcting my mistakes. We worked very hard for about two or three days, and got large pieces of the report written that way. It worked very well.

Neil Armstrong, who was in our group, is extremely good at writing. He would look at my work and immediately find every weak spot, just like that—he was right every time—and I was very impressed.

Each group was writing a chapter or two of the main report. Our group wrote some of the stuff in "Chapter 3: The Accident," but our main work was "Chapter 4: The Cause of the Accident." One result of this system, however, was that we never had a meeting to discuss what each of our groups found out—to comment on each other's findings from our different perspectives. Instead, we did what they call "wordsmithing"—or what Mr. Hotz later called "tomb-stone engraving"—correcting punctuation, refining phrases, and so on. We never had a *real* discussion of ideas, except incidentally in the course of this wordsmithing.

For example, a question would come up: "Should this sentence about the engines be worded this way or that way?"

I would try to get a little discussion started. "From my own experiences, I got the impression that the engines aren't as good as you're saying here . . ."

So they'd say, "Then we'll use the more conservative wording here," and they'd go on to the next sentence.

Perhaps that's a very efficient way to get a report out quickly, but we spent meeting after meeting doing this wordsmithing.

Every once in a while we'd interrupt that to discuss the typography and the color of the cover. And after each discussion, we were asked to vote. I thought it would be most efficient to vote for the same color we had decided on in the meeting before, but it turned out I was always in the minority! We finally chose red. (It came out blue.)

One time I was talking to Sally Ride about something I mentioned in my report on the engines, and she didn't seem to know about it. I said, "Didn't you see my report?"

She says, "No, I didn't get a copy."

So I go over to Keel's office and say, "Sally tells me she didn't get a copy of my report."

He looks surprised, and turns to his secretary. "Please make a copy of Dr. Feynman's report for Dr. Ride."

Then I discover Mr. Acheson hasn't seen it.

"Make a copy and give it to Mr. Acheson."

I finally caught on, so I said, "Dr. Keel, I don't think anybody has seen my report."

So he says to his secretary, "Please make a copy for all the commissioners and give it to them."

Then I said to him, "I appreciate how much work you're doing, and that it's difficult to keep everything in mind. But I thought you told me that you showed my report to everybody."

He says, "Yes, well, I meant all of the staff."

I later discovered, by talking to people on the staff, that they hadn't seen it either.

When the other commissioners finally got to see my report, most of them thought it was very good, and it ought to be in the commission report somewhere.

Encouraged by that, I kept bringing up my report. "I'd

like to have a meeting to discuss what to do with it," I kept saying.

"We'll have a meeting about it next week" was the standard answer. (We were too busy wordsmithing and voting on the color of the cover.)

Gradually I realized that the way my report was written, it would require a lot of wordsmithing—and we were running out of time. Then somebody suggested that my report could go in as an appendix. That way, it wouldn't have to be wordsmithed to fit in with anything else.

But some of the commissioners felt strongly that my report should go in the main report somehow: "The appendices won't come out until months later, so nobody will read your report if it's an appendix," they said.

I thought I'd compromise, however, and let it go in as an appendix.

But now there was a new problem: my report, which I had written on my word processor at home, would have to be converted from the IBM format to the big document system the commission was using. They had a way of doing that with an optical scanning device.

I had to go to a little bit of trouble to find the right guy to do it. Then, it didn't get done right away. When I asked what happened, the guy said he couldn't find the copy I had given him. So I had to give him another copy.

A few days later, I finished writing my report about the avionics, and I wanted to combine it with my report on the engines. So I took the avionics report to the guy and I said, "I'd like to put this in with my other report."

Then I needed to see a copy of my new report for some reason, but the guy gave me an old copy, before the avionics was added. "Where's the new one with the avionics?" I said.

"I can't find it"—and so on. I don't remember all the details, but it seemed my report was always missing or half-cooked. It could easily have been mistakes, but there were too many of them. It was quite a struggle, nursing my report along.

Then, in the last couple of days, when the main report is ready to be sent to the printer, Dr. Keel wants my report to be wordsmithed too, even though it's going in as an appendix. So I took it to the regular editor there, a capable man named Hansen, and he fixed it up without changing the sense of it. Then it was put back into the machine as "Version #23"—there were revisions and revisions.

(By the way: *everything* had 23 versions. It has been noted that computers, which are supposed to increase the speed at which we do things, have not increased the speed at which we write reports: we used to make only three versions—because they're so hard to type—and now we make 23 versions!)

The next day I noticed Keel working on my report: he had put all kinds of big circles around whole sections, with X's through them; there were all kinds of thoughts left out. He explained, "This part doesn't have to go in because it says more or less what we said in the main report."

I tried to explain that it's much easier to get the logic if all the ideas are together, instead of everything being distributed in little pieces all over the main report. "After all," I said, "it's only gonna be an appendix. It won't make any difference if there's a little repetition."

Dr. Keel put back something here and there when I asked him to, but there was still so much missing that my report wasn't anything like it was before.

SOMETIME in May, at one of our last meetings, we got around to making a list of possible recommendations. Somebody would say, "Maybe one of the things we should discuss is the establishment of a safety board."

"Okay, we'll put that down."

I'm thinking, "At last! We're going to have a discussion!"

But it turns out that this tentative list of topics *becomes* the recommendations—that there be a safety board, that there be a this, that there be a that. The only discussion was about which recommendation we should write first, which one should come second, and so forth.

There were many things I wanted to discuss further. For example, in regard to a safety board, one could ask: "Wouldn't such a committee just add another layer to an already overgrown bureaucracy?"

There had been safety boards before. In 1967, after the Apollo accident, the investigating committee at the time invented a special panel for safety. It worked for a while, but it didn't last.

We didn't discuss why the earlier safety boards were no longer effective; instead, we just made up more safety boards: we called them the "Independent Solid Rocket Motor Design Oversight Committee," the "Shuttle Transportation System Safety Advisory

The Tenth Recommendation

Panel," and the "Office of Safety, Reliability, and Quality Assurance." We decided who would oversee each safety board, but we didn't discuss whether the safety boards created by our commission had any better chance of working, whether we could fix the existing boards so they *would* work, or whether we should have them at all.

I'm not as sure about a lot of things as everybody else. Things need to be thought out a little bit, and we weren't doing enough *thinking* together. Quick decisions on important matters are not very good—and at the speed we were going, we were bound to make some impractical recommendations.

We ended up rearranging the list of possible recommendations and wordsmithing them a little, and then we voted yes or no. It was an odd way of doing things, and I wasn't used to it. In fact, I got the feeling we were being railroaded: things were being decided, somehow, a little out of our control.

At any rate, in our last meeting, we agreed to nine recommendations. Many of the commissioners went home after that meeting, but I was going to New York a few days later, so I stayed in Washington.

The next day, I happened to be standing around in Mr. Rogers's office with Neil Armstrong and another commissioner when Rogers says, "I thought we should have a tenth recommendation. Everything in our report is so negative; I think we need something positive at the end to balance it."

He shows me a piece of paper. It says,

The Commission strongly recommends that NASA continue to receive the support of the Administration and the nation. The agency constitutes a national resource and plays a critical role in space exploration and development. It also provides a symbol of national pride and technological leadership. The Commission applauds NASA's spectacular achievements of the past and anticipates impressive achievements to come. The findings and recommendations presented in this report are intended to contribute to the future NASA successes that the nation both expects and requires as the 21st century approaches.

In our four months of work as a commission, we had never discussed a policy question like that, so I felt there was no reason to put it in. And although I'm not saying I disagreed with it, it wasn't obvious that it was true, either. I said, "I think this tenth recommendation is inappropriate."

I think I heard Armstrong say, "Well, if somebody's not in favor of it, I think we shouldn't put it in."

But Rogers kept working on me. We argued back and forth a little bit, but then I had to catch my flight to New York.

While I was in the airplane, I thought about this tenth recommendation some more. I wanted to lay out my arguments carefully on paper, so when I got to my hotel in New York, I wrote Rogers a letter. At the end I wrote, "This recommendation reminds me of the NASA flight reviews: 'There are critical problems, but never mind—keep on flying!'"

It was Saturday, and I wanted Mr. Rogers to read my letter before Monday. So I called up his secretary—everybody was working seven days a week to get the report out in time—and I said, "I'd like to dictate a letter to you; is that all right?"

She says, "Sure! To save you some money, let me call you right back." She calls me back, I dictate the letter, and she hands it directly to Rogers.

When I came back on Monday, Mr. Rogers said, "Dr. Feynman, I've read your letter, and I agree with everything it says. But you've been out-voted."

"Out-voted? How was I out-voted, when there was no meeting?"

Keel was there, too. He says, "We called everybody, and they all agree with the recommendation. They all voted for it."

"I don't think that's fair!" I protested. "If I could have presented my arguments to the other commissioners, I don't think I'd have been out-voted." I didn't know what to do, so I said, "I'd like to make a copy of it."

When I came back, Keel says, "We just remembered that we didn't talk to Hotz about it, because he was in a meeting. We forgot to get his vote."

I didn't know what to make of that, but I found out later that Mr. Hotz was in the building, not far from the copy machine.

Later, I talked to David Acheson about the tenth recommendation. He explained, "It doesn't really mean anything; it's only motherhood and apple pie."

I said, "Well, if it doesn't mean anything, it's not necessary, then."

"If this were a commission for the National Academy of Sciences, your objections would be proper. But don't forget," he says, "this is a presidential commission. We should say something for the President."

"I don't understand the difference," I said. *"Why can't I be careful and scientific when I'm writing a report to the President?"*

Being naive doesn't always work: my argument had no effect. Acheson kept telling me I was making a big thing out of nothing, and I kept saying it weakened our report and it shouldn't go in.

So that's where it ended up: "The Commission strongly recommends that NASA continue to receive the support of the Administration and the nation . . ."—all this "motherhood and apple pie" stuff to "balance" the report.

While I was flying home, I thought to myself, "It's funny that the only part of the report that was *genuinely* balanced was my own report: I said negative things about the engine, and positive things about the avionics. And I had to struggle with them to get it in, even as a lousy appendix!"

I thought about the tenth recommendation. All the other recommendations were based on evidence we had found, but this one had no evidence whatsoever. I could see the whitewash dripping down. It was *obviously* a mistake! It would make our report look bad. I was very disturbed.

When I got home, I talked to Joan, my sister. I told her about the tenth recommendation, and how I had been "out-voted."

"Did you call any of the other commissioners and talk to them yourself?" she said.

"Well, I talked to Acheson, but he was for it."

"Any others?"

"Uh, no." So I called up three other commissioners— I'll call them A, B, and C.

I call A, who says, *"What* tenth recommendation?"

I call B, who says, "Tenth recommendation? What are you talking about?"

I call C, who says, "Don't you remember, you dope? I was in the office when Rogers first told us, and I don't see anything wrong with it."

It appeared that the only people who knew about the tenth recommendation were the people who were in the office when Rogers told us. I didn't bother to make any more telephone calls. After all, it's enough—I didn't feel that I had to open all the safes to check that the combination is the same!*

Then I told Joan about my report—how it was so emasculated, even though it was going in as an appendix.

*This refers to "Safecracker Meets Safecracker," another story told in *Surely You're Joking, Mr. Feynman!*

She says, "Well, if they do that to your report, what have you accomplished, being on the commission? What's the result of all your work?"

"Aha!"

I sent a telegram to Mr. Rogers:

PLEASE TAKE MY SIGNATURE OFF THE REPORT UNLESS TWO THINGS OCCUR: 1) THERE IS NO TENTH RECOMMENDATION, AND 2) MY REPORT APPEARS WITHOUT MODIFICATION FROM VERSION #23.

(I knew by this time I had to define everything carefully.)

In order to get the number of the version I wanted, I called Mr. Hotz, who was in charge of the documentation system and publishing the report. He sent me Version #23, so I had something definite to publish on my own, if worse came to worst.

The result of this telegram was that Rogers and Keel tried to negotiate with me. They asked General Kutyna to be the intermediary, because they knew he was a friend of mine. What a *good* friend of mine he was, they didn't know.

Kutyna says, "Hello, Professor, I just wanted to tell you that I think you're doing very well. But I've been given the job of trying to talk you out of it, so I'm going to give you the arguments."

"Fear not!" I said. "I'm not gonna change my mind. Just give me the arguments, and fear not."

The first argument was that if I don't accept the tenth recommendation, they won't accept my report, even as an appendix.

I didn't worry about that one, because I could always put out my report myself.

All the arguments were like that: none of them was very good, and none of them had any effect. I had thought through carefully what I was doing, so I just stuck to my guns.

Then Kutyna suggested a compromise: they were willing to go along with my report as I wrote it, except for one sentence near the end.

I looked at the sentence and I realized that I had already made my point in the previous paragraph. Repeating the point amounted to polemics; removing the phrase made my report much better. I accepted the compromise.

Then I offered a compromise on the tenth recommendation: "If they want to say something nice about NASA at the end, just don't call it a recommendation, so people will know that it's not in the same class as the other recommendations: call it a 'concluding thought' if you want. And to avoid confusion, don't use the words 'strongly recommends.' Just say 'urges'—'The Commission urges that NASA continue to receive the support of the Administration and the nation.' All the other stuff can stay the same."

A little bit later, Keel calls me up: "Can we say *'strongly urges'*?"

"No. Just 'urges'."

"Okay," he said. And that was the final decision.

Meet the
Press

I PUT my name on the main report, my own report got in as an appendix, and everything was all right. In early June we went back to Washington and gave our report to the President in a ceremony held in the Rose Garden. That was on a Thursday. The report was not to be released to the public until the following Monday, so the President could study it.

Meanwhile, the newspaper reporters were working like demons: they knew our report was finished and they were trying to scoop each other to find out what was in it. I knew they would be calling me up day and night, and I was afraid I would say something about a technical matter that would give them a hint.

Reporters are very clever and persistent. They'll say, "We heard such-and-such—is it true?" And pretty soon, what you're thinking you didn't tell them shows up in the newspaper!

I was determined not to say a word about the report until it was made public, on Monday. A friend of mine convinced me to go on the "MacNeil/Lehrer Newshour," so I said yes for Monday evening's show.

I also had my secretary set up a press conference for Tuesday at Caltech. I said, "Tell the reporters who want to talk to me that I haven't any comment on anything: any questions they have, I'll be glad to answer on

FIGURE 18. *The Commission Report was presented to the president in the Rose Garden at the White House. Visible, from left to right, are General Kutyna, William Rogers, Eugene Covert, President Reagan, Neil Armstrong, and Richard Feynman. (© PETE SOUZA, THE WHITE HOUSE.)*

FIGURE 19. *At the reception. (© PETE SOUZA, THE WHITE HOUSE.)*

Tuesday at my press conference."

Over the weekend, while I was still in Washington, it
leaked somehow that I had threatened to take my name off
the report. Some paper in Miami started it, and soon the
story was running all over about this argument between me
and Rogers. When the reporters who were used to covering
Washington heard "Mr. Feynman has nothing to say; he'll
answer all your questions at his press conference on Tuesday," it sounded suspicious—as though the argument was
still on, and I was going to have this press conference on
Tuesday to explain why I took my name off the report.

But I didn't know anything about it. I isolated myself
from the press so much that I wasn't even reading the
newspapers.

On Sunday night, the commission had a goodbye dinner arranged by Mr. Rogers at some club. After we finished
eating, I said to General Kutyna, "I can't stay around anymore. I have to leave a little early."

He says, "What can be so important?"

I didn't want to say.

He comes outside with me, to see what this "important" something is. It's a bright red sports car with two
beautiful blonds inside, waiting to whisk me away.

I get in the car. We're about to speed off, leaving
General Kutyna standing there scratching his head, when
one of the blonds says, "Oh! General Kutyna! I'm Ms.
So-and-so. I interviewed you on the phone a few weeks
ago."

So he caught on. They were reporters from the "Mac-
Neil/Lehrer Newshour."

They were very nice, and we talked about this and that
for the show Monday night. Somewhere along the line I
told them I was going to have my own press conference on
Tuesday, and I was going to give out my report—even
though it was going to appear as an appendix three months

later. They said my report sounded interesting, and they'd like to see it. By this time we're all very friendly, so I gave them a copy.

They dropped me off at my cousin's house, where I was staying. I told Frances about the show, and how I gave the reporters a copy of my report. Frances puts her hands to her head, horrified.

I said, "Yes, that was a dumb mistake, wasn't it! I'd better call 'em up and tell 'em not to use it."

I could tell by the way Frances shook her head that it wasn't gonna be so easy!

I call one of them up: "I'm sorry, but I made a mistake: I shouldn't have given my report to you, so I'd prefer you didn't use it."

"We're in the news business, Dr. Feynman. The goal of the news business is to get news, and your report is newsworthy. It would be completely against our instincts and practice not to use it."

"I know, but I'm naive about these things. I simply made a mistake. It's not fair to the other reporters who will be at the press conference on Tuesday. After all, would you like it if you came to a press conference and the guy had mistakenly given his report to somebody else? I think you can understand that."

"I'll talk to my colleague and call you back."

Two hours later, they call back—they're both on the line—and they try to explain to me why they should use it: "In the news business, it's customary that whenever we get a document from somebody the way we did from you, it means we can use it."

"I appreciate that there are conventions in the news business, but I don't know anything about these things, so as a courtesy to me, please don't use it."

It went back and forth a little more like that. Then another "We'll call you back," and another long delay. I could tell from the long delays that they were having a lot of trouble with this problem.

I was in a very good fettle, for some reason. I had already lost, and I knew what I needed, so I could focus easily. I had no difficulty admitting complete idiocy—which is usually the case when I deal with the world—and I didn't think there was any law of nature which said I had to give in. I just kept going, and didn't waver at all.

It went late into the night: one o'clock, two o'clock, we're still working on it. "Dr. Feynman, it's very unprofessional to give someone a story and then retract it. This is not the way people behave in Washington."

"It's obvious I don't know anything about Washington. But this is the way *I* behave—like a fool. I'm sorry, but it was simply an error, so as a courtesy, please don't use it."

Then, somewhere along the line, one of them says, "If we go ahead and use your report, does that mean you won't go on the show?"

"You said it; *I* didn't."

"We'll call you back."

Another delay.

Actually, I hadn't decided whether I'd refuse to go on the show, because I kept thinking it was possible I could undo my mistake. When I thought about it, I didn't think I could legitimately play that card. But when one of them made the mistake of proposing the possibility, I said, *"You* said it; *I* didn't"—very cold—as if to say, "I'm not threatening you, but you can figure it out for yourself, honey!"

They called me back, and said they wouldn't use my report.

When I went on the show, I never got the impression that any of the questions were based on my report. Mr. Lehrer did ask me whether there had been any problems between me and Mr. Rogers, but I weaseled: I said there had been no problems.

After the show was over, the two reporters told me

they thought the show went fine without my report. We left good friends.

I flew back to California that night, and had my press conference on Tuesday at Caltech. A large number of reporters came. A few asked questions about my report, but most of them were interested in the rumor that I had threatened to take my name off the commission report. I found myself telling them over and over that I had no problem with Mr. Rogers.

Afterthoughts

NOW that I've had more time to think about it, I still like Mr. Rogers, and I still feel that everything's okay. It's my judgment that he's a fine man. Over the course of the commission I got to appreciate his talents and his abilities, and I have great respect for him. Mr. Rogers has a very good, smooth way about him, so I reserve in my head the possibility—not as a suspicion, but as an unknown—that I like him because he knew how to make me like him. I prefer to assume he's a genuinely fine fellow, and that he is the way he appears. But I was in Washington long enough to know that I can't tell.

I'm not exactly sure what Mr. Rogers thinks of me. He gives me the impression that, in spite of my being such a pain in the ass to him in the beginning, he likes me very much. I may be wrong, but if he feels the way I feel toward him, it's good.

Mr. Rogers, being a lawyer, had a difficult job to run a commission investigating what was essentially a technical question. With Dr. Keel's help, I think the technical part of it was handled well. But it struck me that there were several fishinesses associated with the big cheeses at NASA.

Every time we talked to higher level managers, they kept saying they didn't know anything about the problems below them. We're getting this kind of thing again in the Iran-Contra hearings, but at that time, this kind of situation was new to me: either the guys at the top didn't know, in which case they should have known, or they did know, in which case they're lying to us.

When we learned that Mr. Mulloy had put pressure on Thiokol to launch, we heard time after time that the next level up at NASA knew nothing about it. You'd think Mr. Mulloy would have notified a higher-up during this big discussion, saying something like, "There's a question as to whether we should fly tomorrow morning, and there's been some objection by the Thiokol engineers, but we've decided to fly anyway—what do you think?" But instead, Mulloy said something like, "All the questions have been resolved." There seemed to be some reason why guys at the lower level didn't bring problems up to the next level.

I invented a theory which I have discussed with a considerable number of people, and many people have explained to me why it's wrong. But I don't remember their explanations, so I cannot resist telling you what I think led to this lack of communication in NASA.

When NASA was trying to go to the moon, there was a great deal of enthusiasm: it was a goal everyone was anxious to achieve. They didn't know if they could do it, but they were all working together.

I have this idea because I worked at Los Alamos, and I experienced the tension and the pressure of everybody working together to make the atomic bomb. When somebody's having a problem—say, with the detonator—everybody knows that it's a big problem, they're thinking of ways to beat it, they're making suggestions, and when they hear about the solution they're excited, because that means their work is now useful: if the detonator didn't work, the bomb wouldn't work.

I figured the same thing had gone on at NASA in the early days: if the space suit didn't work, they couldn't go to the moon. So everybody's interested in everybody else's problems.

But then, when the moon project was over, NASA had all these people together: there's a big organization in Houston and a big organization in Huntsville, not to mention at Kennedy, in Florida. You don't want to fire people and send them out in the street when you're done with a big project, so the problem is, what to do?

You have to convince Congress that there exists a project that only NASA can do. In order to do so, it is necessary—at least it was *apparently* necessary in this case—to exaggerate: to exaggerate how economical the shuttle would be, to exaggerate how often it could fly, to exaggerate how safe it would be, to exaggerate the big scientific facts that would be discovered. "The shuttle can make so-and-so many flights and it'll cost such-and-such; we went to the moon, so we can *do* it!"

Meanwhile, I would guess, the engineers at the bottom are saying, "No, no! We can't make that many flights. If we had to make that many flights, it would mean such-and-such!" And, "No, we can't do it for that amount of money, because that would mean we'd have to do thus-and-so!"

Well, the guys who are trying to get Congress to okay their projects don't want to hear such talk. It's better if they don't hear, so they can be more "honest"—they don't want to be in the position of lying to Congress! So pretty soon the attitudes begin to change: information from the bottom which is disagreeable—"We're having a problem with the seals; we should fix it before we fly again"—is suppressed by big cheeses and middle managers who say, "If you tell me about the seals problems, we'll have to ground the shuttle and fix it." Or, "No, no, keep on flying, because otherwise, it'll look bad," or "Don't tell me; I don't want to hear about it."

Maybe they don't say explicitly "Don't tell me," but they discourage communication, which amounts to the same thing. It's not a question of what has been written down, or who should tell what to whom; it's a question of whether, when you *do* tell somebody about some problem, they're *delighted* to hear about it and they say "Tell me more" and "Have you tried such-and-such?" or they say "Well, see what you can do about it"—which is a completely different atmosphere. If you try once or twice to communicate and get pushed back, pretty soon you decide, "To hell with it."

So that's my theory: because of the exaggeration at the top being inconsistent with the reality at the bottom, communication got slowed up and ultimately jammed. That's how it's possible that the higher-ups didn't know.

The other possibility is that the higher-ups did know, and they just *said* they didn't know.

I looked up a former director of NASA—I don't remember his name now—who is the head of some company in California. I thought I'd go and talk to him when I was on one of my breaks at home, and say, "They all *say* they haven't heard. Does that make any sense? How does someone go about investigating them?"

He never returned my calls. Perhaps he didn't want to talk to the commissioner investigating higher-ups; maybe he had had enough of NASA, and didn't want to get involved. And because I was busy with so many other things, I didn't push it.

There were all kinds of questions we didn't investigate. One was this mystery of Mr. Beggs, the former director of NASA who was removed from his job pending an investigation that had nothing to do with the shuttle; he was replaced by Graham shortly before the accident. Nevertheless, it turned out that, every day, Beggs came to his old office. People came in to see him, although he never talked to Graham. What was he doing? Was there some activity still being directed by Beggs?

From time to time I would try to get Mr. Rogers interested in investigating such fishinesses. I said, "We have lawyers on the commission, we have company managers, we have very fine people with a large range of experiences. We have people who know how to get an answer out of a guy when he doesn't want to say something. I don't know how to do that. If a guy tells me the probability of failure is 1 in 10^5, I know he's full of crap—but I don't know what's natural in a bureaucratic system. We oughta get some of the big shots together and ask them questions: just like we asked the second-level managers like Mr. Mulloy, we should ask the first level."

He would say, "Yes, well, I think so."

Mr. Rogers told me later that he wrote a letter to each of the big shots, but they replied that they didn't have anything they wanted to say to us.

There was also the question of pressure from the White House.

It was the President's idea to put a teacher in space, as a symbol of the nation's commitment to education. He had proposed the idea a year before, in his State of the Union address. Now, one year later, the State of the Union speech was coming up again. It would be perfect to have the teacher in space, talking to the President and the Congress. All the circumstantial evidence was very strong.

I talked to a number of people about it, and heard various opinions, but I finally concluded that there was no pressure from the White House.

First of all, the man who pressured Thiokol to change its position, Mr. Mulloy, was a second-level manager. Ahead of time, nobody could predict what might get in the way of a launch. If you imagine Mulloy was told "Make sure the shuttle flies tomorrow, because the President wants it," you'd have to imagine that *everybody else* at his level had to be told—and there are a lot of people at his level. To tell that many people would make it sure to leak out. So that way of putting on pressure was very unlikely.

By the time the commission was over, I understood much better the character of operations in Washington and in NASA. I learned, by seeing how they worked, that the people in a big system like NASA *know* what has to be done—*without* being told.

There was *already* a big pressure to keep the shuttle flying. NASA had a flight schedule they were trying to meet, just to show the capabilities of NASA—never mind whether the president was going to give a speech that night or not. So I don't believe there was any direct activity or any special effort from the White House. There was no need to do it, so I don't believe it was done.

I could give you an analog of that. You know those signs that appear in the back windows of automobiles—those little yellow diamonds that say BABY ON BOARD, and things like that? You don't have to *tell* me there's a baby on board; I'm gonna drive carefully *anyway!* What am I supposed to do when I see there's a baby on board: act differently? As if I'm suddenly gonna drive more carefully and not hit the car because there's a baby on board, when all I'm trying to do is not hit it anyway!

So NASA was trying to get the shuttle up anyway: you don't have to say there's a baby on board, or there's a teacher on board, or it's important to get this one up for the President.

Now that I've talked to some people about my experiences on the commission, I think I understand a few things that I didn't understand so well earlier. One of them has to do with what I said to Dr. Keel that upset him so much. Recently I was talking to a man who spent a lot of time in Washington, and I asked him a particular question which, if he didn't take it right, could be considered a grave insult. I would like to explain the question, because it seems to me to be a real possibility of what I said to Dr. Keel.

The only way to have real success in science, the field

I'm familiar with, is to describe the evidence very carefully without regard to the way you feel it should be. If you have a theory, you must try to explain what's good and what's bad about it equally. In science, you learn a kind of standard integrity and honesty.

In other fields, such as business, it's different. For example, almost every advertisement you see is obviously designed, in some way or another, to fool the customer: the print that they don't want you to read is small; the statements are written in an obscure way. It is obvious to anybody that the product is not being presented in a scientific and balanced way. Therefore, in the selling business, there's a lack of integrity.

My father had the spirit and integrity of a scientist, but he was a salesman. I remember asking him the question "How can a man of integrity be a salesman?"

He said to me, "Frankly, many salesmen in the business are not straightforward—they think it's a better way to sell. But I've tried being straightforward, and I find it has its advantages. In fact, I wouldn't do it any other way. If the customer thinks at all, he'll realize he has had some bad experience with another salesman, but hasn't had that kind of experience with you. So in the end, several customers will stay with you for a long time and appreciate it."

My father was not a big, successful, famous salesman; he was the sales manager for a medium-sized uniform company. He was successful, but not enormously so.

When I see a congressman giving his opinion on something, I always wonder if it represents his *real* opinion or if it represents an opinion that he's designed in order to be elected. It seems to be a central problem for politicians. So I often wonder: what is the relation of integrity to working in the government?

Now, Dr. Keel started out by telling me that he had a degree in physics. I always assume that everybody in physics has integrity—perhaps I'm naive about that—so I must have asked him a question I often think about: "How can a man of integrity get along in Washington?"

It's very easy to read that question another way: "Since you're getting along in Washington, you can't be a man of integrity!"

Another thing I understand better now has to do with where the idea came from that cold affects the O-rings. It was General Kutyna who called me up and said, "I was working on my carburetor, and I was thinking: what is the effect of cold on the O-rings?"

Well, it turns out that one of NASA's own astronauts told him there was information, somewhere in the works of NASA, that the O-rings had no resilience whatever at low temperatures—and NASA wasn't saying anything about it.

But General Kutyna had the career of that astronaut to worry about, so the *real* question the General was thinking about while he was working on his carburetor was, "How can I get this information out without jeopardizing my astronaut friend?" His solution was to get the professor excited about it, and his plan worked perfectly.

Introduction

Appendix F: Personal Observations on the Reliability of the Shuttle

It appears that there are enormous differences of opinion as to the probability of a failure with loss of vehicle and of human life.* The estimates range from roughly 1 in 100 to 1 in 100,000. The higher figures come from working engineers, and the very low figures come from management. What are the causes and consequences of this lack of agreement? Since 1 part in 100,000 would imply that one could launch a shuttle each day for 300 years expecting to lose only one, we could properly ask, "What is the cause of management's fantastic faith in the machinery?"

We have also found that certification criteria used in flight readiness reviews often develop a gradually decreasing strictness. The argument that the same risk was flown before without failure is often accepted as an argument for the safety of accepting it again. Because of this, obvious weaknesses are accepted again and again—sometimes without a sufficiently serious attempt to remedy them, sometimes without a flight delay because of their continued presence.

There are several sources of infor-

*Leighton's note: The version printed as Appendix F in the commission report does not appear to have been edited, so I took it upon myself to smooth it out a little bit.

mation: there are published criteria for certification, including a history of modifications in the form of waivers and deviations; in addition, the records of the flight readiness reviews for each flight document the arguments used to accept the risks of the flight. Information was obtained from direct testimony and reports of the range safety officer, Louis J. Ullian, with respect to the history of success of solid fuel rockets. There was a further study by him (as chairman of the Launch Abort Safety Panel, LASP) in an attempt to determine the risks involved in possible accidents leading to radioactive contamination from attempting to fly a plutonium power supply (called a radioactive thermal generator, or RTG) on future planetary missions. The NASA study of the same question is also available. For the history of the space shuttle main engines, interviews with management and engineers at Marshall, and informal interviews with engineers at Rocketdyne, were made. An independent (Caltech) mechanical engineer who consulted for NASA about engines was also interviewed informally. A visit to Johnson was made to gather information on the reliability of the avionics (computers, sensors, and effectors). Finally, there is the report "A Review of Certification Practices Potentially Applicable to Man-rated Reusable Rocket Engines," prepared at the Jet Propulsion Laboratory by N. Moore et al. in February 1986 for NASA Headquarters, Office of Space Flight. It deals with the methods used by the FAA and the military to certify their gas turbine and rocket engines. These authors were also interviewed informally.

Solid Rocket Boosters (SRB)

An estimate of the reliability of solid-fuel rocket boosters (SRBs) was made by the range safety officer by studying the experience of all previous rocket flights. Out of a total of nearly 2900 flights, 121 failed (1 in 25). This includes, however, what may be called "early errors"—rockets flown for the first few times in which design errors are discovered and fixed. A more reasonable figure for the mature rockets might be 1 in 50. With special care in selecting parts and in inspection, a figure below 1 in 100 might be achieved, but 1 in 1000 is probably not attainable with today's technology. (Since there are two rockets on the shuttle, these rocket failure rates must be doubled to get shuttle failure rates due to SRB failure.)

NASA officials argue that the figure is much lower. They point out that "since the shuttle is a manned vehicle, the probability of mission success is necessarily very close to 1.0." It is not very clear what this phrase means. Does it mean it *is* close to 1 or that it *ought to be* close to 1? They go on to explain, "Historically, this extremely high degree of mission success has given rise to a difference in philosophy between manned space flight programs and unmanned programs; i.e., numerical probability usage versus engineering judgment." (These quotations are from "Space Shuttle Data for Planetary Mission RTG Safety Analysis," pages 3-1 and 3-2, February 15, 1985, NASA, JSC.) It is true that if the probability of failure was as low as 1 in 100,000 it would take an inordinate number of tests to determine it: you would get nothing but a string of perfect flights with no precise figure—other than that the probability is likely less than the number of such flights in the string so far. But if the real probability is not so small, flights would show troubles, near failures, and possibly actual failures with a reasonable number of trials, and standard statistical methods could give a reasonable estimate. In fact, previous NASA experience had shown, on occasion, just such difficulties, near accidents, and even accidents, all giving warning that the probability of flight failure was not so very small.

Another inconsistency in the argument not to determine reliability through historical experience (as the range safety officer did) is NASA's appeal to history: "Historically, this high degree of mission success . . ." Finally, if we are to replace standard numerical probability usage with engineering judgment, why do we find such an enormous disparity between the management estimate and the judgment of the engineers? It would appear that, for whatever purpose—be it for internal or external consumption—the management of NASA exaggerates the reliability of its product to the point of fantasy.

The history of the certification and flight readiness reviews will not be repeated here (see other parts of the commission report), but the phenomenon of accepting seals that had shown erosion and blowby in previous flights is very clear. The *Challenger* flight is an excellent example: there are several references to previous flights; the acceptance and success of these flights are taken as evidence of safety. But erosion and blowby are not what the design expected. They are warnings that something is wrong. The equipment is not operating as expected, and therefore there is a danger that it can operate with even wider deviations in this unexpected and not thoroughly understood way. The fact that this danger did not lead to a catastrophe before is no guarantee that it will not the next time, unless it is completely understood. When playing Russian roulette, the fact that the first shot got off safely is of little comfort for the next. The origin and consequences of the erosion and blowby were not understood. Erosion and blowby did not occur equally on all flights or in all joints: sometimes there was more, sometimes less. Why not sometime, when whatever conditions determined it were right, wouldn't there be still more, leading to catastrophe?

In spite of these variations from case to case, officials behaved as if they understood them, giving apparently logical arguments to each other—often citing the "success" of previous flights. For example, in determining if flight 51-L was safe to fly in the face of ring erosion in flight 51-C, it was noted that the erosion depth was only one-third of the radius. It had been noted in an experiment cutting the ring that cutting it as deep as one radius was necessary before the ring failed. Instead of being very concerned that variations of poorly understood conditions might reasonably create a deeper erosion this time, it was asserted there was "a safety factor of three."

This is a strange use of the engineer's term "safety factor." If a bridge is built to withstand a certain load without the beams permanently deforming, cracking, or breaking, it may be designed for the materials used to actually stand up under three times the load. This "safety factor" is to allow for uncertain excesses of load, or unknown extra loads, or weaknesses in the material that might have unexpected flaws, et cetera. But if the expected load comes on to the new bridge and a crack appears in a beam, this is a failure of the design. There was no safety factor at all, even though the bridge did not actually collapse because the crack only went one-third of the way through the beam. The O-rings of the solid rocket boosters were not designed to erode. Erosion was a clue that something was wrong. Erosion was not something from which safety could be inferred.

There was no way, without full understanding, that one could have confidence that conditions the next time might not produce erosion three times more severe than the time before. Nevertheless, officials fooled themselves into thinking they had such understanding and confidence, in spite of the peculiar variations from case to case. A mathematical model was made to calculate erosion. This was a model based not on physical understanding but on empirical curve fitting. Specifically, it was supposed that a stream of hot gas impinged on the O-ring material, and the heat was determined at the point of stagnation (so far, with reasonable physical, thermodynamical laws). But to determine how much rubber eroded, it was assumed that the erosion varied as the .58 power of heat, the .58 being determined by a nearest fit. At any rate, adjusting some other numbers, it was determined that the model agreed with the erosion (to a depth of one-third the radius of the ring). There is nothing so wrong with this analysis as believing the answer! Uncertainties appear everywhere in the model. How strong the gas stream might be was unpredictable; it depended on holes formed in the putty. Blowby showed that the ring might fail, even though it was only partially eroded. The empirical formula was known to be uncertain, for the curve did not go directly through the very data points by which it was determined. There was a cloud of points, some twice above and some twice below the fitted curve, so erosions twice those predicted were reasonable from that cause alone. Similar uncertainties surrounded the other constants in the formula, et cetera, et cetera. When using a mathematical model, careful attention must be given to the uncertainties in the model.

Space Shuttle Main Engines (SSME)

During the flight of the 51-L the three space shuttle main engines all worked perfectly, even beginning to shut down in the last moments as the fuel supply began to fail. The question arises, however, as to whether—had the engines failed, and we were to investigate them in as much detail as we did the solid rocket booster—we would find a similar lack of attention to faults and deteriorating safety criteria. In other words, were the organization weaknesses that contributed to the accident confined to the solid rocket booster sector, or were they a more general characteristic of NASA? To that end the space shuttle main engines and the avionics were both investigated. No similar study of the orbiter or the external tank was made.

The engine is a much more complicated structure than the solid rocket booster, and a great deal more detailed engineering goes into it. Generally, the engineering seems to be of high quality, and apparently considerable attention is paid to deficiencies and faults found in engine operation.

The usual way that such engines are designed (for military or civilian aircraft) may be called the component system, or bottom-up design. First it is necessary to thoroughly understand the properties and limitations of the materials to be used (turbine blades, for example), and tests are begun in experimental rigs to determine those. With this knowledge, larger component parts (such as bearings) are designed and tested individually. As deficiencies and design errors are noted they are corrected and verified with further testing. Since one tests only parts at a time, these tests and modifications are not overly expensive. Finally one works up to the final design of the entire engine, to the necessary specifications. There is a good chance, by this time, that the engine will generally succeed, or that any failures are easily isolated and analyzed because the failure modes, limitations of materials, et cetera, are so well understood. There is a very good chance that the modifications to get around final difficulties in the engine are not very hard to make, for most of the serious problems have already been discovered and dealt with in the earlier, less expensive stages of the process.

The space shuttle main engine was handled in a different manner—top down, we might say. The engine was designed and put together all at once with relatively little detailed preliminary study of the materials and components. But now, when troubles are found in bearings, turbine blades, coolant pipes, et cetera, it is more expensive and difficult to discover the causes and make changes. For example, cracks have been found in the turbine blades of the high-pressure oxygen turbopump. Are they caused by flaws in the material, the effect of the oxygen atmosphere on the properties of the material, the thermal stresses of startup or shutdown, the vibration and stresses of steady running, or mainly at some resonance at certain speeds, or something else? How long can we run from crack initiation to crack failure, and how does this depend on power level? Using the completed engine as a test bed to resolve such questions is extremely expensive. One does not wish to lose entire engines in order to find out where and how failure occurs. Yet, an accurate knowledge of this information is essential to acquiring a confidence in the engine reliability in use. Without detailed understanding, confidence cannot be attained.

A further disadvantage of the top-down method is that if an understanding of a fault is obtained, a simple fix—such as a new shape for the turbine housing—may be impossible to implement without a redesign of the entire engine.

The space shuttle main engine is a very remarkable machine. It has a greater ratio of thrust to weight than any previous engine. It is built at the edge of—sometimes outside of—previous engineering experience. Therefore, as expected, many different kinds of flaws and difficulties have turned up. Because, unfortunately, it was built in a top-down manner, the flaws are difficult to find and to fix. The design aim of an engine lifetime of 55 mission equivalents (27,000 seconds of operation, either in missions of 500 seconds each or on a test stand) has not been obtained. The engine now requires very frequent maintenance and replacement of important parts such as turbopumps, bearings, sheet metal housings, et cetera. The high-pressure fuel turbopump had to be replaced every three or four mission equivalents (although this may have been fixed, now) and the high-pressure oxygen turbopump every five or six. This was, at most, 10 percent of the original design specifications. But our main concern here is the determination of reliability.

In a total of 250,000 seconds of operation, the main

engines have failed seriously perhaps 16 times. Engineers pay close attention to these failings and try to remedy them as quickly as possible by test studies on special rigs experimentally designed for the flaw in question, by careful inspection of the engine for suggestive clues (like cracks), and by considerable study and analysis. In this way, in spite of the difficulties of top-down design, through hard work many of the problems have apparently been solved.

A list of some of the problems (and their status) follows:

Turbine blade cracks in high-pressure fuel turbopumps (HPFTP). (May have been solved.)

Turbine blade cracks in high-pressure oxygen fuel turbopumps (HPOTP). (Not solved.)

Augmented spark igniter (ASI) line rupture. (Probably solved.)

Purge check valve failure. (Probably solved.)

ASI chamber erosion. (Probably solved.)

HPFTP turbine sheet metal cracking. (Probably solved.)

HPFTP coolant liner failure. (Probably solved.)

Main combustion chamber outlet elbow failure. (Probably solved.)

Main combustion chamber inlet elbow weld offset. (Probably solved.)

HPOTP subsynchronous whirl. (Probably solved.)

Flight acceleration safety cutoff system (partial failure in a redundant system). (Probably solved.)

Bearing spalling. (Partially solved.)

A vibration at 4000 hertz making some engines inoperable. (Not solved.)

Many of these apparently solved problems were the early difficulties of a new design: 13 of them occurred in the first 125,000 seconds and only 3 in the second 125,000 seconds. Naturally, one can never be sure that all the bugs are out; for some, the fix may not have addressed the true cause. Thus it is not unreasonable to guess there may be at least one surprise in the next 250,000 seconds, a probability of $1/500$ per engine per mission. On a mission there are three engines, but it is possible that some accidents would be self-contained and affect only one engine. (The shuttle can abort its mission with only two engines.) Therefore, let us say that the unknown surprises do not, in and of themselves, permit us to guess that the probability of mission failure due to the space shuttle main engines is less than $1/500$. To this we must add the chance of failure from known, but as yet unsolved, problems. These we discuss below.

(Engineers at Rocketdyne, the manufacturer, estimate the total probability as $1/10,000$. Engineers at Marshall estimate it as $1/300$, while NASA management, to whom these engineers report, claims it is $1/100,000$. An independent engineer consulting for NASA thought 1 or 2 per 100 a reasonable estimate.)

The history of the certification principles for these engines is confusing and difficult to explain. Initially the rule seems to have been that two sample engines must each have had twice the time operating without failure, as the operating time of the engine to be certified (rule of $2x$). At least that is the FAA practice, and NASA seems to have adopted it originally, expecting the certified time to be 10 missions (hence 20 missions for each sample). Obviously, the best engines to use for comparison would be those of greatest total operating time (flight plus test), the so-called fleet leaders. But what if a third sample engine and several others fail in a short time? Surely we will not be safe because two were unusual in lasting longer. The short time might be more representative of the real possibilities, and in the spirit of the safety factor of 2, we should only operate at half the time of the short-lived samples.

The slow shift toward a decreasing safety factor can be seen in many examples. We take that of the HPFTP turbine blades. First of all the idea of testing an entire engine was abandoned. Each engine has had many important parts (such as the turbopumps themselves) replaced at frequent intervals, so the rule of 2x must be shifted from engines to components. Thus we accept an HPFTP for a given certification time if two samples have each run successfully for twice that time (and, of course, as a practical matter, no longer insisting that this time be as long as 10 missions). But what is "successfully"? The FAA calls a turbine blade crack a failure, in order to really provide a safety factor greater than 2 in practice. There is some time that an engine can run between the time a crack originally starts and the time it has grown large enough to fracture. (The FAA is contemplating new rules that take this extra safety time into account, but will accept them only if it is very carefully analyzed through known models within a known range of experience and with materials thoroughly tested. None of these conditions applies to the space shuttle main engines.)

Cracks were found in many second-stage HPFTP turbine blades. In one case three were found after 1900 seconds, while in another they were not found after 4200 seconds, although usually these longer runs showed cracks. To follow this story further we must realize that the stress depends a great deal on the power level. The *Challenger* flight, as well as previous flights, was at a level called 104 percent of rated power during most of the time the engines were operating. Judging from some material data, it is supposed that at 104 percent of rated power, the time to crack is about twice that at 109 percent, or full power level (FPL). Future flights were to be at 109 percent because of heavier payloads, and many tests were made at this level. Therefore, dividing time at 104 percent by 2, we obtain units called equivalent full power level (EFPL). (Obviously, some uncertainty is introduced by that, but it has not been studied.) The earliest cracks mentioned above occurred at 1375 seconds EFPL.

Now the certification rule becomes "limit all second-stage blades to a maximum of 1375 seconds EFPL." If one objects that the safety factor of 2 is lost, it is pointed out that the one turbine ran for 3800 seconds EFPL without cracks, and half of this is 1900 so we are being more conservative. We have fooled ourselves in three ways. First, we have only one sample, and it is not the fleet leader: the other two samples of 3800 or more seconds EFPL had 17 cracked blades between them. (There are 59 blades in the engine.) Next, we have abandoned the 2x rule and substituted equal time (1375). And finally, the 1375 is where a crack was discovered. We can say that no crack had been found below 1375, but the last time we looked and saw no cracks was 1100 seconds EFPL. We do not know when the crack formed between these times. For example, cracks may have been formed at 1150 seconds EFPL. (Approximately two-thirds of the blade sets tested in excess of 1375 seconds EFPL had cracks. Some recent experiments have, indeed, shown cracks as early as 1150 seconds.) It was important to keep the number high, for the shuttle had to fly its engines very close to their limit by the time the flight was over.

Finally, it is claimed that the criteria have not been abandoned, and that the system is safe, by giving up the FAA convention that there should be no cracks, and by considering only a completely fractured blade a failure. With this definition no engine has yet failed. The idea is that since there is sufficient time for a crack to grow to fracture, we can ensure that all is safe by inspecting all blades for cracks. If cracks are found, replace the blades, and if none are found, we have enough time for a safe mission. Thus, it is claimed, the crack problem is no longer a flight safety problem, but merely a maintenance problem.

This may in fact be true. But how well do we know that

cracks always grow slowly enough so that no fracture can occur in a mission? Three engines have run for long time periods with a few cracked blades (about 3000 seconds EFPL), with no blade actually breaking off.

A fix for this cracking may have been found. By changing the blade shape, shot-peening the surface, and covering it with insulation to exclude thermal shock, the new blades have not cracked so far.

A similar story appears in the history of certification of the HPOTP, but we shall not give the details here.

In summary, it is evident that the flight readiness reviews and certification rules show a deterioration in regard to some of the problems of the space shuttle main engines that is closely analogous to the deterioration seen in the rules for the solid rocket boosters.

Avionics

By "avionics" is meant the computer system on the orbiter as well as its input sensors and output actuators. At first we will restrict ourselves to the computers proper, and not be concerned with the reliability of the input information from the sensors of temperature, pressure, et cetera; nor with whether the computer output is faithfully followed by the actuators of rocket firings, mechanical controls, displays to astronauts, et cetera.

The computing system is very elaborate, having over 250,000 lines of code. Among many other things it is responsible for the automatic control of the shuttle's entire ascent into orbit, and for the descent until the shuttle is well into the atmosphere (below Mach 1), once one button is pushed deciding the landing site desired. It would be possible to make the entire landing automatic. (The landing gear lowering signal is expressly left out of computer control, and must be provided by the pilot, ostensibly for safety reasons.) During orbital flight the computing system is used in the control of payloads, in the display of information to the astronauts, and in the exchange of information with the ground. It is evident that the safety of flight requires guaranteed accuracy of this elaborate system of computer hardware and software.

In brief, hardware reliability is ensured by having four essentially independent identical computer systems. Where possible, each sensor also has multiple copies— usually four—and each copy feeds all four of the computer lines. If the inputs from the sensors disagree, either a certain average or a majority selection is used as the effective input, depending on the circumstances. Since each computer sees all copies of the sensors, the inputs are the same, and because the algorithms used by each of the four computers are the same, the results in each computer should be identical at each step. From time to time they are compared, but because they might operate at slightly different speeds, a system of stopping and waiting at specified times is instituted before each comparison is made. If one of the computers disagrees or is too late in having its answer ready, the three which do agree are assumed to be correct and the errant computer is taken completely out of the system. If, now, another computer fails, as judged by the agreement of the other two, it is taken out of the system, and the rest of the flight is canceled: descent to the landing site is instituted, controlled by the two remaining computers. It is seen that this is a redundant system since the failure of only one computer does not affect the mission. Finally, as an extra feature of safety, there is a fifth independent computer, whose memory is loaded with only the programs for ascent and descent, and which is capable of controlling the descent if there is a failure of more than two of the computers of the main line of four.

There is not enough room in the memory of the main-line computers for all the programs of ascent, descent, and payload programs in flight, so the memory is loaded by the astronauts about four times from tapes.

Because of the enormous effort required to replace the software for such an elaborate system and to check out a new system, no change in the hardware has been made since the shuttle transportation system began about fifteen years ago. The actual hardware is obsolete—for example, the memories are of the old ferrite-core type. It is becoming more difficult to find manufacturers to supply such old-fashioned computers that are reliable and of high enough quality. Modern computers are much more reliable, and they run much faster. This simplifies circuits and allows more to be done. Today's computers would not require so much loading from tapes, for their memories are much larger.

The software is checked very carefully in a bottom-up fashion. First, each new line of code is checked; then sections of code (modules) with special functions are verified. The scope is increased step by step until the new changes are incorporated into a complete system and checked. This complete output is considered the final product, newly released. But working completely independently is a verification group that takes an adversary attitude to the software development group and tests the software as if it were a customer of the delivered product. There is additional verification in using the new programs in simulators, et cetera. An error during this stage of verification testing is considered very serious, and its origin is studied very carefully to avoid such mistakes in the future. Such inexperienced errors have been found only about six times in all the programming and program changing (for new or altered payloads) that has been done. The principle followed is: all this verification is not an aspect of program safety; it is a test of that safety in a noncatastrophic verification. Flight safety is to be judged solely on how well the programs do in the verified tests. A failure here generates considerable concern.

To summarize, then, the computer software checking system is of highest quality. There appears to be no process of gradually fooling oneself while degrading standards, the process so characteristic of the solid rocket booster and space shuttle main engine safety systems. To be sure, there have been recent suggestions by management to curtail such elaborate and expensive tests as being unnecessary at this late date in shuttle history. Such suggestions must be resisted, for they do not appreciate the mutual subtle influences and sources of error generated by even small program changes in one part of a program on another. There are perpetual requests for program changes as new payloads and new demands and modifications are suggested by the users. Changes are expensive because they require extensive testing. The proper way to save money is to curtail the number of requested changes, not the quality of testing for each.

One might add that the elaborate system could be very much improved by modern hardware and programming techniques. Any outside competition would have all the advantages of starting over. Whether modern hardware is a good idea for NASA should be carefully considered now.

Finally, returning to the sensors and actuators of the avionics system, we find that the attitude toward system failure and reliability is not nearly as good as for the computer system. For example, a difficulty was found with certain temperature sensors sometimes failing. Yet eighteen months later the same sensors were still being used, still sometimes failing, until a launch had to be scrubbed because two of them failed at the same time. Even on a succeeding flight this unreliable sensor was used again. And reaction control systems, the rocket jets used for reorienting and control in flight, still are somewhat unreliable. There is considerable redundancy, but also a long history of failures, none of which has yet been extensive enough to seriously affect a flight. The action of the jets is checked by sensors: if a jet fails to fire, the computers choose another jet to fire. But they are not designed to fail, and the problem should be solved.

Conclusions

If a reasonable launch schedule is to be maintained, engineering often cannot be done fast enough to keep up with the expectations of the originally conservative certification criteria designed to guarantee a very safe vehicle. In such situations, safety criteria are altered subtly—and with often apparently logical arguments—so that flights can still be certified in time. The shuttle therefore flies in a relatively unsafe condition, with a chance of failure on the order of a percent. (It is difficult to be more accurate.)

Official management, on the other hand, claims to believe the probability of failure is a thousand times less. One reason for this may be an attempt to assure the government of NASA's perfection and success in order to ensure the supply of funds. The other may be that they sincerely believe it to be true, demonstrating an almost incredible lack of communication between the managers and their working engineers.

In any event, this has had very unfortunate consequences, the most serious of which is to encourage ordinary citizens to fly in such a dangerous machine—as if it had attained the safety of an ordinary airliner. The astronauts, like test pilots, should know their risks, and we honor them for their courage. Who can doubt that McAuliffe* was equally a person of great courage, who was closer to an awareness of the true risks than NASA management would have us believe?

Let us make recommendations to ensure that NASA

*Note for foreign readers: Christa McAuliffe, a schoolteacher, was to have been the first ordinary citizen in space—a symbol of the nation's commitment to education, and of the shuttle's safety.

officials deal in a world of reality, understanding technological weaknesses and imperfections well enough to be actively trying to eliminate them. They must live in a world of reality in comparing the costs and utility of the shuttle to other methods of entering space. And they must be realistic in making contracts and in estimating the costs and difficulties of each project. Only realistic flight schedules should be proposed—schedules that have a reasonable chance of being met. If in this way the government would not support NASA, then so be it. NASA owes it to the citizens from whom it asks support to be frank, honest, and informative, so that these citizens can make the wisest decisions for the use of their limited resources.

For a successful technology, reality must take precedence over public relations, for Nature cannot be fooled.

WHEN I was younger, I thought science would make good things for everybody. It was obviously useful; it was good. During the war I worked on the atomic bomb. This result of science was obviously a very serious matter: it represented the destruction of people.

After the war I was very worried about the bomb. I didn't know what the future was going to look like, and I certainly wasn't anywhere near sure that we would last until now. Therefore one question was—is there some evil involved in science?

Put another way—what is the value of the science I had dedicated myself to—the thing I loved—when I saw what terrible things it could do? It was a question I had to answer.

"The Value of Science" is a kind of report, if you will, on many of the thoughts that came to me when I tried to answer that question.

Richard Feynman

The Value of Science*

FROM time to time people suggest to me that scientists ought to give more consideration to social problems—especially that they should be more responsible in considering the impact of science on society. It seems to be generally believed that if the scientists would only look at these very difficult social problems and not spend so much time fooling with less vital scientific ones, great success would come of it.

It seems to me that we *do* think about these problems from time to time, but we don't put a full-time effort into them—the reasons being that we know we don't have any magic formula for solving social problems, that social problems are very much harder than scientific ones, and that we usually don't get anywhere when we do think about them.

I believe that a scientist looking at nonscientific problems is just as dumb as the next guy—and when he talks about a nonscientific matter, he sounds as naive as anyone untrained in the matter. Since the question of the value of science is *not* a scientific subject, this talk is dedicated to proving my point—by example.

The first way in which science is of value is familiar to everyone. It is that scientific knowledge enables us to do all kinds of things and to make all kinds

*A public address given at the 1955 autumn meeting of the National Academy of Sciences.

of things. Of course if we make *good* things, it is not only to the credit of science; it is also to the credit of the moral choice which led us to good work. Scientific knowledge is an enabling power to do either good or bad—but it does not carry instructions on how to use it. Such power has evident value—even though the power may be negated by what one does with it.

I learned a way of expressing this common human problem on a trip to Honolulu. In a Buddhist temple there, the man in charge explained a little bit about the Buddhist religion for tourists, and then ended his talk by telling them he had something to say to them that they would *never* forget—and I have never forgotten it. It was a proverb of the Buddhist religion:

To every man is given the key to the gates of heaven; the same key opens the gates of hell.

What then, is the value of the key to heaven? It is true that if we lack clear instructions that enable us to determine which is the gate to heaven and which the gate to hell, the key may be a dangerous object to use.

But the key obviously has value: how can we enter heaven without it?

Instructions would be of no value without the key. So it is evident that, in spite of the fact that it could produce enormous horror in the world, science is of value because it *can* produce *something*.

Another value of science is the fun called intellectual enjoyment which some people get from reading and learning and thinking about it, and which others get from working in it. This is an important point, one which is not considered enough by those who tell us it is our social responsibility to reflect on the impact of science on society.

Is this mere personal enjoyment of value to society as a whole? No! But it is also a responsibility to consider the aim of society itself. Is it to arrange matters so that people can enjoy things? If so, then the enjoyment of science is as important as anything else.

But I would like *not* to underestimate the value of the world view which is the result of scientific effort. We have been led to imagine all sorts of things infinitely more marvelous than the imaginings of poets and dreamers of the past. It shows that the imagination of nature is far, far greater than the imagination of man. For instance, how much more remarkable it is for us all to be stuck—half of us upside down—by a mysterious attraction to a spinning ball that has been swinging in space for billions of years than to be carried on the back of an elephant supported on a tortoise swimming in a bottomless sea.

I have thought about these things so many times alone that I hope you will excuse me if I remind you of this type of thought that I am sure many of you have had, which no one could ever have had in the past because people then didn't have the information we have about the world today.

For instance, I stand at the seashore, alone, and start to think.

> There are the rushing waves
> mountains of molecules
> each stupidly minding its own business
> trillions apart
> yet forming white surf in unison.
>
> Ages on ages
> before any eyes could see
> year after year
> thunderously pounding the shore as now.
> For whom, for what?
> On a dead planet
> with no life to entertain.
>
> Never at rest
> tortured by energy
> wasted prodigiously by the sun

poured into space.
A mite makes the sea roar.

Deep in the sea
all molecules repeat
the patterns of one another
till complex new ones are formed.
They make others like themselves
and a new dance starts.

Growing in size and complexity
living things
masses of atoms
DNA, protein
dancing a pattern ever more intricate.

Out of the cradle
onto dry land
here it is
standing:
atoms with consciousness;
matter with curiosity.

Stands at the sea,
wonders at wondering: I
a universe of atoms
an atom in the universe.

The same thrill, the same awe and mystery, comes
again and again when we look at any question deeply
enough. With more knowledge comes a deeper, more wonderful mystery, luring one on to penetrate deeper still.
Never concerned that the answer may prove disappointing,
with pleasure and confidence we turn over each new stone
to find unimagined strangeness leading on to more wonderful questions and mysteries—certainly a grand adventure!

It is true that few unscientific people have this particular type of religious experience. Our poets do not write
about it; our artists do not try to portray this remarkable

thing. I don't know why. Is no one inspired by our present picture of the universe? This value of science remains unsung by singers: you are reduced to hearing not a song or poem, but an evening lecture about it. This is not yet a scientific age.

Perhaps one of the reasons for this silence is that you have to know how to read the music. For instance, the scientific article may say, "The radioactive phosphorus content of the cerebrum of the rat decreases to one-half in a period of two weeks." Now what does that mean?

It means that phosphorus that is in the brain of a rat—and also in mine, and yours—is not the same phosphorus as it was two weeks ago. It means the atoms that are in the brain are being replaced: the ones that were there before have gone away.

So what is this mind of ours: what are these atoms with consciousness? Last week's potatoes! They now can *remember* what was going on in my mind a year ago—a mind which has long ago been replaced.

To note that the thing I call my individuality is only a pattern or dance, *that* is what it means when one discovers how long it takes for the atoms of the brain to be replaced by other atoms. The atoms come into my brain, dance a dance, and then go out—there are always new atoms, but always doing the same dance, remembering what the dance was yesterday.

When we read about this in the newspaper, it says "Scientists say this discovery may have importance in the search for a cure for cancer." The paper is only interested in the use of the idea, not the idea itself. Hardly anyone can understand the importance of an idea, it is so remarkable. Except that, possibly, some children catch on. And when a child catches on to an idea like that, we have a scientist. It is too late* for them to get the spirit when they are in our

---

*I would now say, "It is late—although not too late—for them to get the spirit . . ."

universities, so we must attempt to explain these ideas to children.

I would now like to turn to a third value that science has. It is a little less direct, but not much. The scientist has a lot of experience with ignorance and doubt and uncertainty, and this experience is of very great importance, I think. When a scientist doesn't know the answer to a problem, he is ignorant. When he has a hunch as to what the result is, he is uncertain. And when he is pretty darn sure of what the result is going to be, he is still in some doubt. We have found it of paramount importance that in order to progress we must recognize our ignorance and leave room for doubt. Scientific knowledge is a body of statements of varying degrees of certainty—some most unsure, some nearly sure, but none *absolutely* certain.

Now, we scientists are used to this, and we take it for granted that it is perfectly consistent to be unsure, that it is possible to live and *not* know. But I don't know whether everyone realizes this is true. Our freedom to doubt was born out of a struggle against authority in the early days of science. It was a very deep and strong struggle: permit us to question—to doubt—to not be sure. I think that it is important that we do not forget this struggle and thus perhaps lose what we have gained. Herein lies a responsibility to society.

We are all sad when we think of the wondrous potentialities human beings seem to have, as contrasted with their small accomplishments. Again and again people have thought that we could do much better. Those of the past saw in the nightmare of their times a dream for the future. We, of *their* future, see that their dreams, in certain ways surpassed, have in many ways remained dreams. The hopes for the future today are, in good share, those of yesterday.

It was once thought that the possibilities people had were not developed because most of the people were ignorant. With universal education, could all men be Voltaires?

Bad can be taught at least as efficiently as good. Education is a strong force, but for either good or evil.

Communications between nations must promote understanding—so went another dream. But the machines of communication can be manipulated. What is communicated can be truth or lie. Communication is a strong force, but also for either good or evil.

The applied sciences should free men of material problems at least. Medicine controls diseases. And the record here seems all to the good. Yet there are some patiently working today to create great plagues and poisons for use in warfare tomorrow.

Nearly everyone dislikes war. Our dream today is peace. In peace, man can develop best the enormous possibilities he seems to have. But maybe future men will find that peace, too, can be good and bad. Perhaps peaceful men will drink out of boredom. Then perhaps drink will become the great problem which seems to keep man from getting all he thinks he should out of his abilities.

Clearly, peace is a great force—as are sobriety, material power, communication, education, honesty, and the ideals of many dreamers. We have more of these forces to control than did the ancients. And maybe we are doing a little better than most of them could do. But what we ought to be able to do seems gigantic compared with our confused accomplishments.

Why is this? Why can't we conquer ourselves?

Because we find that even great forces and abilities do not seem to carry with them clear instructions on how to use them. As an example, the great accumulation of understanding as to how the physical world behaves only convinces one that this behavior seems to have a kind of meaninglessness. The sciences do not directly teach good and bad.

Through all ages of our past, people have tried to fathom the meaning of life. They have realized that if some direction or meaning could be given to our actions, great human forces would be unleashed. So, very many answers have been given to the question of the meaning of it all. But the answers have been of all different sorts, and the proponents of one answer have looked with horror at the actions of the believers in another—horror, because from a disagreeing point of view all the great potentialities of the race are channeled into a false and confining blind alley. In fact, it is from the history of the enormous monstrosities created by false belief that philosophers have realized the apparently infinite and wondrous capacities of human beings. The dream is to find the open channel.

What, then, is the meaning of it all? What can we say to dispel the mystery of existence?

If we take everything into account—not only what the ancients knew, but all of what we know today that they didn't know—then I think we must frankly admit that *we do not know*.

But, in admitting this, we have probably found the open channel.

This is not a new idea; this is the idea of the age of reason. This is the philosophy that guided the men who made the democracy that we live under. The idea that no one really knew how to run a government led to the idea that we should arrange a system by which new ideas could be developed, tried out, and tossed out if necessary, with more new ideas brought in—a trial-and-error system. This method was a result of the fact that science was already showing itself to be a successful venture at the end of the eighteenth century. Even then it was clear to socially minded people that the openness of possibilities was an opportunity, and that doubt and discussion were essential to progress into the unknown. If we want to solve a problem that we have never solved before, we must leave the door to the unknown ajar.

We are at the very beginning of time for the human

race. It is not unreasonable that we grapple with problems. But there are tens of thousands of years in the future. Our responsibility is to do what we can, learn what we can, improve the solutions, and pass them on. It is our responsibility to leave the people of the future a free hand. In the impetuous youth of humanity, we can make grave errors that can stunt our growth for a long time. This we will do if we say we have the answers now, so young and ignorant as we are. If we suppress all discussion, all criticism, proclaiming "This is the answer, my friends; man is saved!" we will doom humanity for a long time to the chains of authority, confined to the limits of our present imagination. It has been done so many times before.

It is our responsibility as scientists, knowing the great progress which comes from a satisfactory philosophy of ignorance, the great progress which is the fruit of freedom of thought, to proclaim the value of this freedom; to teach how doubt is not to be feared but welcomed and discussed; and to demand this freedom as our duty to all coming generations.

Index

A special one-hour audio-cassette tape (and CD) of Richard Feynman playing drums and telling his most infamous story has been produced by Ralph Leighton and Tohru Ohnuki. "Richard Feynman: Safecracker Suite" can be ordered by sending a cheque for £8 (£12 for CD) to: Ralph Leighton, Box 70021, Pasadena CA 91117 USA. All proceeds go to UCLA's John Wayne Cancer Clinic, whose doctors gave Feynman six additional years of life – and gave the rest of us six additional years of Richard Feynman.